Noesis

NOESIS

Essays in the history and philosophy of science, philosophy of language, epistemology and political philosophy

edited by

Stephen Rainey
and Barbara Gabriella Renzi

Cambridge Scholars Press

Noesis: Essays in the history and philosophy of science, philosophy of langugage, epistemology and political philosophy.
Edited by Stephen Rainey and Barbara Gabriella Renzi

This book first published 2005 by

Cambridge Scholars Press

15 Angerton Gardens, Newcastle, NE5 2JA, UK

British Library Cataloguing in Publication Data
A catalogue record for this book is available from the British Library

ISBN 1-904303-49-8

CONTENTS

FOREWORD

This volume presents the papers given at an international conference of postgraduate students in Queen's University Belfast in April 2004. The conference was the third in an annual series "Universities of Ireland Postgraduate Philosophy Conference".

The Belfast conference attracted about forty participants from Universities all over Ireland and the UK and also from the USA and Canada and several other countries in Europe. Roy Cook (then in the *Arche* project at St Andrews, now in a post at Villanova University, Pennsylvania) was the main invited speaker, and he was joined by twelve further speakers. Eight of the papers delivered are reproduced, following suitable reflection and revision, in this volume.

Senior philosophers in Queen's lent strong support, in academic and financial terms, to this conference. In a sense we were repaying this cohort of our postgraduate students for the key role which they played in facilitating the highly successful major Belfast philosophy conferences in July 2003 – the Joint Session of the Mind Association & Aristotelian Society, the British Society for Ethical Theory, and the British Society for the Philosophy of Science. 260 philosophers came to Belfast at that time, and their appreciation of this experience has resonated ever since. The success of that event depended crucially on the hands-on work of eight of our students during the week of the conferences.

The students in turn took strong inspiration from this experience of the senior philosophical world, and in this way they were motivated to organise the event which this volume celebrates. The diachronic interaction between the conferences themselves and the personnel who sustained them presents something of a model for postgraduate training in philosophy. We are all being required to engage in such training; and there is some pain in discovering what this requirement involves. The Belfast experience through 2003-2004 may provide exemplary help here.

After several decades of occupancy of the Chair of Logic & Metaphysics at Queen's University Belfast, I reflect increasingly on the nodal position of Belfast in the world-wide connections of philosophy and philosophers. We are integrated into the UK structure of academic institutions; and we also enjoy intimate connections with organisations in the Republic of Ireland, notably the Royal Irish Academy, as well as all higher education institutions. Then there is the significance of Ireland as a non-English entry point for philosophers from Europe and the continent of America. All of this gives Belfast a unique position, which has contributed to the success of these conferences.

I am most grateful to the Cambridge Scholars Press for undertaking the publication of this work, and of course to Barbara Gabriella Renzi and Stephen Rainey for their editorial labours in bringing this enterprise to fruition.

Prof. David Evans

PREFACE

In April 2004, the Queen's University Philosophy Society and Postgraduate Seminar Group hosted a two day conference open to postgraduates in which students could showcase their latest work in an informed and critical environment. In previous years, this same kind of conference had been held in Dublin and so this was a first not just for Queen's, but for Northern Ireland.

Many papers presented over the two days went on to be published in noteworthy journals, an indication of the high quality on display. Given the number of papers presented, not all could be presented in this volume, but what follows is a selection of some of the best received papers, which cover areas as diverse as philosophy of science and mind, epistemology, political philosophy and ethics and a time span stretching from the classical era to modern analytic philosophy. With such a diverse group of philosophical minds gathered together, the chance to hear a new take on a familiar field, or to hear discussion of areas of philosophy one had never come across was not wasted. Each paper was eagerly received and hopefully all present came away with new ideas and perspectives.

The diversity of the subject matter under investigation among these papers is the strength of this collection, as it introduces the reader to the vast range of important issues still pressing in philosophy today. The pieces are notable for the accessible way in which they get across effectively their often complex subject matter.

What came across most from the conference itself was the value not only of listening to unfamiliar perspectives on familiar topics, or even listening to insightful work in areas not previously studied, but simply in getting together in an atmosphere of eager inquiry with like-minded people. If nothing else, we proved that as much can be learned over a pint or a cup of coffee in the right company as can be in a lecture room.

The conference would not have been possible without the gracious help of the School of Philosophical Studies at Queen's. Special thanks is due to Professor David Evans, Dr. Alan Weir, Mr. Christopher McKnight and Dr. George Pavlakos whose attendance at several of the papers, and chairing of various sessions was much appreciated. Thanks also to Mary Emmerson, secretary of the School of Philosophical Studies, whose help and advice was invaluable.

The Editors

CHAPTER ONE: EMMA TOBIN

WHAT MAKES THE SPECIAL SCIENCES SPECIAL? EXPLORING SCIENTIFIC METHODOLOGY IN THE SPECIAL SCIENCES.

The disunity of the sciences hypothesis introduces a distinction between the physical sciences like physics and chemistry and the special sciences such as geology, biology, psychology and the various medical sciences. These later special sciences are considered 'special' in so far as they cannot be straightforwardly reduced to physics. Reduction is prohibited by the empirical hypothesis that special science kinds are multiply realisable in physical kinds, known as the multiple realisability thesis (henceforth MRT). In this paper, I will assess how philosophers have proceeded to cash out the notion of a special science given this distinction between the special and physical sciences. I will argue that there have been oversights in how one proceeds to assess special sciences assuming the merits of MRT.

The most serious issue, I will argue, is the assumption that a special science must satisfy all of the stringent positivistic criteria set forth by successful science (i.e. in the hard physical sciences like physics), such as explanatory rigour, wide scope, testability, projectibility and universal generalisations (laws). I will be arguing for the view that there is a basic difference between the kinds of the special and physical sciences. As a result, the criteria used to assess the scientific legitimacy of the special sciences ought to be toto coelo different from those that we use to assess the hard physical sciences. More particularly, in assessing the role of putative ceteris paribus laws and the natural kinds that they refer to in special science explanations, I will argue that the criteria for what makes a special science special are quite different to what has been assumed.

The Disunity of The Special Sciences

MRT claims that it is an open empirical possibility that the kind predicate(s) of a reduced science (i.e. a special science) may correspond to a widely disjunctive and heterogeneous set of kind predicate(s) in a reducing science (i.e. physics) and hence that there is no one-to-one correlation between the predicates of physics and those of the special sciences. Fodor provides the following example:

Gresham's Law says something about what will happen in monetary exchanges under certain conditions. I am willing to believe that physics is general in the sense that it implies that any event which consists of a monetary exchange (hence any event which falls under Gresham's law) has a true description in the vocabulary of physics and in virtue of which it falls under the laws of physics. But banal considerations suggest that a physical description which covers all such events must be wildly disjunctive. Some monetary exchanges involve strings of wampum. Some involve dollar bills. And some involve signing one's name to a check. What are the chances that a disjunction of physical predicates which covers all these events (i.e. a disjunctive predicate which can form the right hand side of a bridge law of the form 'x is a monetary exchange ↔ …..') expresses a physical kind? (Fodor (1981) : 134)[1] [Italics Fodor's]

Therefore, the economic kind 'monetary exchange' is multiply realisable in the heterogeneous set of physical kinds that could be counted as examples of monetary exchanges.

Rosenberg (1985) provides another excellent example of the failure of the reduction of a special science (Mendelian genetics) to a lower-level science (molecular genetics), in so far as it would be impossible to reduce the terms of one theory to those of another. He states:

Such vast, unwieldy, general statements as we might construct - in which a Mendelian gene is equated with a molecular one - will be full of disjunctions, conjunctions, negations, exceptions, qualifications. It will make so many appeals to stages in the pathway between the DNA and the phenotypic endpoint of the pathway that it will be without any scientific standing. (Rosenberg (1985) : 107).

What does MRT mean for how we think about the stratification of the sciences? One standard approach that most people will be familiar with is to argue that MRT motivates a demarcation between the special sciences and the physical ones. In this vein, Fodor (1974) (1991) takes the empirical hypothesis of MRT seriously and then uses it to argue that the special sciences are scientifically legitimate, in so far as there are genuine empirical special science laws that refer to real natural kinds. So, realism about the special sciences does not automatically follow from MRT, but one also needs an account of why the special sciences are scientifically legitimate

---

[1] The article was originally published as Fodor, J. (1974) 'Special sciences, or the disunity of science as a working hypothesis' *Synthese*, 28:97-115. All references to the article in this paper will be to the reprinted version of the article in Fodor (1981).

themselves. Fodor (1974) illustrates his account in the following way (see Fig 1 below):

Fig 1:

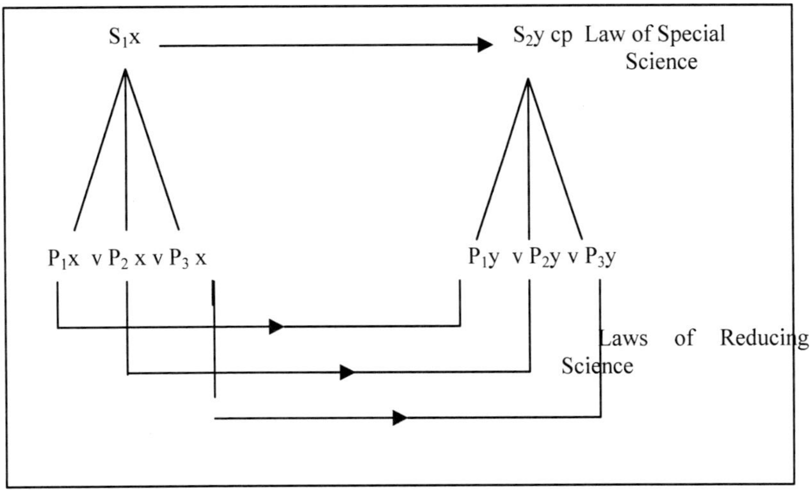

The conditional in (1) below is a genuine law of the special science (S), which is not exceptionless.

$S_1x \rightarrow S_2y$ cp

Where (1) should be read as 'all events which consist of x's being S_1 bring about events which consist of y's being S_2 all else being equal' . Both S_1x and S_2y pick out real special science natural kinds. Were the paradigm of reductionism to be correct then we should have a candidate physical law, which should in principle reduce the special science law (1) above. A conditional like (2) below would express such a physical law of physics (P).

$P_1x \rightarrow P_1y$[2].

[2] According to Fodor, the connective in the conditional must be transitive. This is important because it is not literally presumed that the special science will reduce to physics directly, but rather that the special science proceeds via bridge laws which connect their predicates with those of intermediary theories. Thus, psychology will reduce to neurology, which will reduce to biochemistry, which will reduce to physics. (See Fodor (1981) :129)

Where (2) should be read as 'all events which consist of x's being P_1 bring about events which consist of y's being P_2, where P_1 and P_2 are predicates of physics and (2) is a physical law. Moreover, if the special science (S) were reducible to physics (P) then it should be possible to reduce laws like (1) above to laws like (2) above via biconditional bridge principles like (3) & (4) below:

$S_1x \leftrightarrow P_1x$[3]
$S_2y \leftrightarrow P_1y$

So therefore, (3) and (4) above should be reducible to (2) above.

However, Fodor argues that no special science kind such as S_1x is coextensive with one physical kind predicate (P_1x) where some bridge principle like (3) expresses a nomologically necessary property identity. For Fodor, laws of the special science like (1) above express interesting counterfactual supporting generalisations about special science kinds whose physical descriptions have absolutely nothing in common[4]. Hence, laws like (1) above are irreducible to the heterogeneous group of realisers that may implement them.

An example will prove helpful here. In intentional psychology a putative law will be something like the following:

$\forall$ (x) (p) [(x fears that p) $\supset$ (x desires that ~p) ceteris paribus][5]

The idea is that the special science law generalises about various properties that might be disparate at the micro-physiological level. So, if we

---

[3] Of course, bridge laws should hold by virtue of the identity of the events that satisfy their antecedents, because Fodor is a token physicalist rather than a type physicalist.

[4] Special science generalisations are considered 'interesting' according to Fodor, because often whether the physical kinds that implement the special science kinds on different occasions have anything in common is entirely irrelevant to the truth of the generalisation, its degree of confirmation, its testability, its projectibility and so on.

[5] Fodor takes this argument to generalise to other special sciences also. Take the putative law in economics that 'A high minimum wage increases unemployment *cp*'. A high minimum wage makes labour intensive processes more expensive and so unemployment will increase because staff will be a more expensive business commodity. However, we can imagine cases where this does not occur, for example, if the wages of employees are raised, then employees will be able to afford to purchase more goods and so the demand for goods will increase and since an increase in demand will require an increase in supply then more workers will be needed. Therefore, we can see that "A high minimum wage increases unemployment" is only true 'all other things being equal' or 'in ideal circumstances'.

look at all the neurophysiological states that implement different tokens of fear in different human beings, or even different tokens of fear in the same human being over time, there will be many candidate realising states ($P_1x \vee P_2x \vee P_3x$). (Moreover, $(\forall x)(\forall y)[(P_1x \vee P_2x \vee P_3x)_{\rightarrow}(P_1y \vee P_2y \vee P_3y)]$ cannot be considered a genuine law because the antecedent and the consequent so formulated will be wildly disjunctive. So, the open generalisations (i.e. laws with disjunctives) fail to pick out the truth of higher-order generalisations. Fodor (1997) explains this in greater detail:

For, to offer a law of the form ($R_1 \vee R_2 \vee$ .....) $\rightarrow$ Q is to invite the charge that one has failed to correctly identify the property in virtue of which the antecedent of the law necessitates the consequent. Or to say the same thing the other way around: Someone who offers such a law undertakes the burden to provide a positive reason that there isn't a higher level but nondisjunctive property of things that are $R_1 \vee R_2$....in virtue of which they bring it about that Q. (Fodor (1997): 18-19.)

Therefore, the burden of proof is on the reductionist to explain how macro-level natural kinds are to be locally reduced to their micro-level disjunctive properties[6]. In order to do this, Fodor argues that we have first of all to attack the generalisations that those natural kinds feature in. In other words, in order to show that there are no self-standing natural kinds in the special sciences, we have to make a compelling case for why there are no genuine special science laws[7]. Importantly, for Fodor the kind predicates of a

---

[6] (Kim (1993) : 10) argues that reduction is not prohibited by the multiple realisability thesis and that disunificationism is consequently an incoherent claim. Although, there cannot be global bridge principles that connect special science laws with the reducing scientific laws that implement them, we can specify bridge laws that are local and context-specific. Therefore, according to Kim, we can give a perfectly coherent account of disjunctive heterogeneous kinds that operate in laws. However, Fodor's (1998) reply is that the burden of proof is on the reductionist to explain why the special sciences do not have self-standing empirical laws. I will return to this in section III.

[7] Fodor sees no problem in making a case for special science laws. He maintains that there has been an error in thinking that laws only cover causal relations. For Fodor, they also function as the expression of non-causal higher-order generalisations that can only be viewed from that level. Fodor maintains that a law may be essential for the purposes of explanation and prediction, despite the fact that the predicates contained in the law are strictly speaking causally superfluous. He states:

> What's gone wrong, I think is failing to notice that laws do more than cover singular causal relations[.....]the function of laws in explaining and predicting events that fall under them need not be its *whole* function. A law may be essential for capturing generalistations even where overdetermination would in principle make it superfluous for causal explanation. When this is the case we

science are nothing more than "the bound variables of its proper laws." (Fodor (1981) : 132). Therefore, Fodor makes the kinds of the special sciences parasitic on the laws. Once a case for ceteris paribus special science laws can be made, Fodor is content to reify special science kinds and grant these sciences autonomy.

However, why the special sciences should require the particular characteristics (i.e. laws and natural kinds) attributed to them by Fodor is an assumption that is built into the understanding of the honorific of "legitimate science". In other words, Fodor assumes that the special sciences should fulfil just the same criteria as their physical science counterpart, in order for them to be considered 'scientific'. The problem here is that background assumptions about the criteria that are constitutive of successful science make theorists like Fodor anxious about classifying the special sciences as anything less than (or indeed different from) their physical science counterparts.

Consequently, Fodor thinks that there are only two routes for him to take. He can reject MRT and the distinction between the physical and special sciences that it motivates (i.e. deny that there are any special sciences), or he can painstakingly give an account of how the special sciences fulfil the positivistic template for successful science, such as providing an account of empirical laws in the special sciences. Fodor argues that the latter account can be provided. Nevertheless, interestingly enough he does admit that he is building into his account a rather murky construal of the relationship between laws and kinds.

If I knew what a law was, and I believed that scientific theories consist just of bodies of laws then I would say [.....] roughly, the kind predicates of a science are the ones whose terms are the bound variables in its proper laws. I am inclined to say that even in my present state of ignorance, accepting the consequence that it makes the murky notion of a kind viciously dependent on the equally murky notion of law and theory. There is no firm footing here. (Fodor (1981) : 132)

I wish to argue that the reason behind why Fodor builds in this murky notion of laws and kinds is because he wishes to deny the latter horn of the following false dilemma for the special sciences.

Either special sciences have no ontological independence from lower-level sciences or, worse, they have no empirical laws, which is just to say that they are not empirical sciences at all. (Shapiro (2000) : 160)

I contend that this is a false disjunction because it hails from an over-stringent policing of the language of science, which assumes that the only alternative to controlled experiments that test universally quantified laws in

---

put up with the overdetermination because we want the generalisation.' (Fodor (1991) : 30)

an objective way, is the description (rather than explanation) of accidental and contingent generalisations.

I wish to argue that there are further methodological questions that need to be answered before a judgement can be made regarding how we ought to classify the special sciences, even if we cannot make a case for special science empirical laws. In other words, the issue is whether the criteria that we use to assess the physical sciences capture the phenomena that support the inductive inferences we make in the special sciences. In particular, I will focus on the following interrelated issues, namely, whether the special sciences require laws, natural kinds, and covering-law explanations to be deemed legitimate.

Special Science Laws

The special sciences observe special science kinds and make general hypotheses about the regularities that hold between them. These generalisations of the special sciences are marked by heterogeneity and exceptions. For this reason, they are often construed as ceteris paribus laws. What are putative ceteris paribus laws? The term ceteris paribus is a Latin term that originated in the 17th century. It can be interpreted in many ways. The standard interpretation is as ‘all other things being equal”. However, it may also be construed as “all other things being constant”, or “operating on the ideal assumption that nothing else changes”.

In physics, to say that something holds ceteris paribus is to operate on the idealised assumption that all background considerations remain the same. For example, most laws assume that the law of inertia (Newton’s First Law) holds, namely, that things tend to remain either at motion or at rest. Equally, they would assume that the conservation laws (i.e. of mass and energy) are in place[8]. Specific laws describe ideal experimental situations. Another example is Boyle’s gas law, which states that if the temperature of a gas is constant, then the product of pressure and volume must remain constant. The reason why this law might only be thought to hold ceteris paribus is because it assumes that the gas is tested within normal densities and pressures. There are exceptions to the law such as when a real gas is compressed to an excessively high density.

[8] In physics, these laws are ones that have the widest scope and application. For this reason, they are often called the most fundamental laws of nature. An example is Lavoisier’s law of the conservation of mass, which states that in a closed system, mass always remains constant, no matter what processes occur internal to the system.

In a similar vein, it is argued that the generalisations of the special sciences ought to be construed as ceteris paribus laws. Examples of putative cp law's are Mendel's laws in genetics. Mendel's first law, the law of segregation states:

(6) In normal Mendelian segregation, at a genetic locus, half of an organism's offspring inherit one of the alleles and the other half the other allele cp. (Stern & Sherwood (1966))[9] [Italics added]

The term 'normal' is used here because there are exceptional cases in which the law fails to hold. In these cases one of the alleles is consistently found in more than half the offspring. Mendel experiments consisted of interbreeding pea plants in order to observe the traits (characteristics) that would appear in subsequent generations. He noticed that when he crossbred plants with round peas and those with wrinkled peas, all of the resulting offspring had round peas. He concluded that if an organism has two different alleles corresponding to the same genetic trait, then one of the two would be dominant (R). So, while both alleles might be passed on to the subsequent generations, the trait of the recessive allelle (r) will not be expressed if only passed on by one parent, since the trait passed on by the dominant allele will dominate. Consequently, the trait of the dominant allele will have a selective advantage, since evolution acts on traits or phenotypes. Of course, this itself may not present a serious problem. In fact, Mendel himself supplemented the original generalisation with one about dominant/recessive genes. Mendel's law of dominance states that:

(7) In normal Mendelian segregation, where one allele is dominant, all of an organism's offspring inherit the dominant allele. [Italics added]

There are exceptions to these laws. For example, in cases of incomplete dominance, the inheritance of a dominant and a usually recessive allele results in a blending of traits to produce intermediate characteristics. For example, in snapdragons incomplete dominance of the traits occurs when red and white are crossed. The offspring are pink, therefore blending the two traits.[10] Another exception to Mendelian genetics involves genes with multiple alleles. Certain traits are controlled by multiple alleles that have complex rules of dominance. Examples are the alleles for human skin and

---

[9] From a collection of translated articles and letters in Stern, C., and E. R. Sherwood. 1966. The Origin of Genetics: A Mendel Source Book. New York: W. H. Freeman.

[10] Plants with red flowers have two copies of the dominant allele R for red flower colour (RR). Plants with white flowers have two copies of the recessive allele r for white flower colour (rr). Pink flowers result in plants with one copy of each allele (Rr), with each allele contributing to a blending of colours.

blood groups[11]. Another exception to Mendel's law is the case of gene linkage. Sometimes when alleles are situated close to one another on a chromosome they produce offspring with unexpected allele combinations. This is because of a process called 'crossing-over'[12]. An example of an exception to this rule is the fruit fly.[13]

Clearly then, Mendel's laws only hold given the proviso that some genetic variation does not occur and given that a system of sexual reproduction is in place in the first place, and so on. So, it is more correct to say that Mendel's law of segregation only holds cp, i.e. in the absence of disturbing factors (where the proviso should list all the possible exceptions that might occur). When Mendel's law is classified as a universal generalisation, then it is strictly speaking false. However, by providing the cp proviso laws with exceptions can be given the same logical structure as those laws that furnish us with explanations in physics; namely, those laws that are thought to provide the template for which generalisations ought to be considered as genuine 'laws'. The cp proviso screens off the disturbing factors, so that science can show why the law holds in the idealised setting when "all else is equal."

There are genuine problems in the construal of putative ceteris paribus laws. Firstly, there is a semantic problem in providing truth conditions for cp laws. It can be argued that cp clauses have no clearly determinate content in so far as "all else being equal" refers to an indeterminate set of affairs. Consequently, since any putative cp law will have no determinate empirical content, it can be argued that it is vacuous (Schiffer (1991), Earman & Roberts (1999)). Secondly, there is a methodological problem in so far as it is not clear how these cp clauses can be tested. Since, very often, we do not know what it means for all else to be equal in advance of testing any given cp law, so how then can it ever be confirmed? Clearly, there is no finite set of empirical observations that can verify such claims. Even if such claims were testable, how could we know that the tests are legitimate, since scientists can count just about anything as an exception so as to retain the legitimacy of the law? Therefore, analysing the special sciences in terms of ceteris paribus laws might not be as helpful as was originally thought, in so far as the generalisations of the special sciences will be considered vacuous or

[11] Karl Landsteiner first discovered blood groups with complex multiple alleles.

[12] This is a process during meiosis when two chromosomes (situated close together) may interconnect and exchange places on the chromosome. The pair then breaks apart to form two chromosomes with different genes.

[13] T.H. Morgan (1910) discovered the fruit fly *(Drosophila)*. Drosophila eyes are normally bright red. Morgan discovered an exceptional white-eyed male.

untestable and worse still unfalsifiable. Clearly, the cp locution is unhelpful here.

However, I wish to pose the prior question as to why we require a construal of special science generalisations as laws in the first place? In the next sections, I will show how a failure to cash out precisely the relationship between laws and natural kinds has resulted in misleading claims about the disunity of science. More particularly, I will reject the claim that realism about the special sciences requires an account of genuine cp laws. Secondly, I will show how the assumption that laws are required for genuine scientific explanations fuels the cp account. In rejecting these assumptions about the relationship between laws and natural kinds and special science explanations we see that cp laws need not strictly speaking be required. In fact, a glimpse at the special sciences in practice reveals that the criteria for what makes a special science special are altogether different to what is otherwise assumed.

CP Laws and Natural Kinds

As outlined above, according to Fodor, the special sciences have genuine empirical laws (qualified cp), which refer to real natural kinds. In his earlier work, Fodor (1974), (1981) leaves the relationship between laws and kinds rather unexplained. He accepts a weak construal of special science kinds, such that the multiply realised kinds of a special science are no more than "the bound variables in its proper laws". (Fodor (1981) : 132) The relationship between laws and kinds is a rather murky one. Fodor is making the laws of the special sciences more basic than the natural kind predicates that figure in these laws. However, I wish to argue that an argument for the disunity of the sciences cannot afford to leave this relationship murky. In fact, how we are to understand cp laws in the special sciences (and indeed whether we should even require them) depends on this very relationship.

Determining the exact relationship between laws and natural kinds is clearly problematic. It is important to consider whether laws of nature or natural kinds are indeed more basic. In the end, the commitments we make here will determine how we should understand the disunity of the sciences given MRT. On the one hand, we want the laws of a theory to be those that allow us to make inductive inferences to future instances of the kind predicates expressed in the law. We want the laws somehow to corroborate our inductive claims. On the other hand, in so far as we want to be able to make stable inductive inferences, we want the necessity we understand laws to have to be corroborated by the actual ontological relations in the world; be it relations between universals, essential properties or whatever. Moreover, this ontological account is important in so far as we want to be realists even in a limited sense about scientific knowledge. It is rather difficult to pin down precisely whether it is laws or kinds that are more basic, that is to say, which one, if either, can provide some ontological ground to underpin the claims we make to scientific explanation in science more generally. Moreover, whatever account of the special sciences that we provide should at least be consistent with how science proceeds in scientific practice.

Kim (1992) has argued that special science kinds are not themselves nomic kinds. In fact, they can be locally reduced to their disjunctive realisers. So, special science kinds are purely conceptual, in so far as future science should provide an account of these realising kinds. Kim agrees with theorists like Fodor that multiple realisability makes reduction ipso facto impossible. However, he argues that local reductions are possible. Take the example of Jade. We now know, contrary to what was once believed, that jade is not a mineral kind, rather it is composed of Jadeite and Nephrite. We can rightly say that Jade is multiply realisable (henceforth MR) in Jadeite and Nephrite. Kim argues that since Jade is paradigmatically MR, then it is unprojectible and so a proposition like:

Jade is green

cannot be a law. Firstly, genuine laws are projectible; in so far as observation of positive instances of F's that are G's increases our belief that the next F will be a G. Kim imagines the theoretical possibility that all the past positive instance of (8) were really instances of Jadeite. So (9) below is confirmed rather than (8):

(9) Jadeite is green.

So, we cannot say that what we thought in the past to be positive instances really confirmed (8). So, (8) is not confirmed by its instances and so is not a law. Kim would argue then that there are no laws about Jade and that Jade itself is not a genuine mineral kind. Of course, Kim takes this to generalise to other special science laws and indeed to generalise to the view that there are no 'autonomous' special sciences. Therefore, the ipso facto failure to reduce special science kinds should not lead to the disunity of the sciences. He states:

> We put the following question to Fodor and like-minded philosophers: if pain is nomically equivalent to N, the property claimed to be wildly disjunctive and obviously nonnomic, why isn't pain itself equally heterogeneous and nonnomic as a kind? (Kim (1993) : 323).

Fodor's (1998 : 12) first response is that Kim has not really shown that (8) is unprojectible, but rather makes the methodological point that in scientific practice we would realise that we have in fact made a sampling error, is so far as our past testing involved a biased sample (i.e. all samples of Jade were in fact Jadeite). Nevertheless, Fodor admits that Kim's challenge does present a problem for his account of cp laws. Although green Jadeite samples tell us that Jadeite is green, they tell us nothing about the colour of nephrite and nephrite samples tell us that nephrite is green, but they tell us nothing about the colour of Jadeite. For this reason:

Jadeite is green iff nephrite is green

although true, is a purely accidental generalisation. In a similar vein, we could say that (11) below is of the same ilk.

(11) Jadeite is green iff grass is green.

It is in response to this article that Fodor (1998) pins down his view on the relationship between laws and kinds. Fodor introduces, by way of response, a distinction between a multiply based property that is disjunctive (MD) and a multiply based property that is disjunctively realised (MDR). I wish to formulate the distinction in terms of the following two biconditionals:

(MD): Ψ is a multiply based property that is disjunctive iff there is no possible world W_2 where there is some property β that realises Ψ, which is not a property that realises Ψ in the actual world W_1.

(MDR): Ψ is a multiply based property that is disjunctively realised iff there is some metaphysically possible world W_2, where there is some property β that realises Ψ, which is not a property, which realises Ψ in the actual world W_1.

Jade is an instance of MD rather than MDR, because there is no instance of Jade in any metaphysically possible world where it has any realizer that it lacks in the actual world (i.e. there is no possible world where we have an instance of Jade, which is not realised by either nephrite or jadeite). So, something with all the other qualitative properties of Jade in another possible world but which lacked the essential properties (i.e. Jadeite or Nephrite) would simple not be an instance of Jade. In contrast, pain is an instance of MDR because we can imagine some metaphysically possible world where a token pain state is realised by silicon chips or z-fibres firing, or whatever. It was this broader understanding of multiple realisability which was to preclude all forms of reduction (even local reductions)[14]. Presumably, Fodor

[14] The whole point of functionalism in the philosophy of mind was that the multiple realisability thesis is to be understood in this broader sense.

would take it that the other special sciences are also 'autonomous' as a result of MDR. Therefore, the disunity of the special sciences is motivated by MDR understood as a trans-global claim.

Importantly, Fodor is arguing here that the kinds of the special sciences should not be understood as kinds defined by their essential properties. Kinds that are (MD) are understood in this way because if you want to have Jade it must be composed essentially of either Jadeite or Nephrite[15]. In other words, if J (Jade) is realised by the disjunction P (Jadeite) or (Q) Nephrite, then it is metaphysically necessary that the properties a thing has qua J are either properties it has qua P or properties it has qua Q. In contrast, kinds that are (MRD) are defined by their functional causal roles rather than by the essential properties of the kinds that realise them. So, it does not matter if pains are realised by silicon chips, neurons, c-fibres or whatever, rather it matters that pains obey a certain causal role. Importantly, this causal role is revealed in the propositions expressed in cp special science laws. For example, one aspect of the functional role of pain is expressed in the following proposition:

(12) Pain causes avoidance cp.

'Pain' is not understood in terms of some essential properties of natural kinds that make any given instance of pain an instance of pain; rather, it is understood as the kind of mental state that functions to cause pain avoidance behaviour.

The reason why the special sciences remain 'autonomous' is not only that there cannot be any reduction of special science kinds to the disjunctive kinds that realise them at the physical level, but also because it is the special science kind which is the bound variable in the cp law, (whatever way it is realised), that is projectible. Fodor makes the further methodological point that if there is a higher-level property that subsumes all the states that satisfy an open disjunction, then we will want to formulate our laws in terms of it. So, even if there is no reason for us to reject laws that have disjunctions as such, then methodologically we ought to, because the special science generalisation (law) is stronger and is confirmed by more evidence than the corresponding disjunction. Therefore, if special science laws are legitimate in so far as they are confirmed by their instances and support and corroborate the causal-explanatory claims of the special sciences, then this is enough to reify the properties of special science kinds. He states:

Why are we prepared to buy nondisjunctive laws at the cost of reifying high-level properties? My story is that this policy complies with a dictate that inductive practice obeys quite generally: Prefer the strongest claim

[15] This is the same as the water = H_2O case in Putnam.

compatible with the evidence, all else being equal. Quantification over instances is one aspect of rational compliance with this injunction; reification of higher level kinds is another. (Fodor (1998) : 20; emphasis Fodor's)

Therefore, if the evidence supports the generalisations that we make in the special sciences, then we should treat of those generalisations as laws of nature. So, we can see that Fodor is making two claims here. Firstly, the generalisations of the special sciences are projectible, in so far as they are confirmed by their instances and so should be analysed as laws of nature. Secondly, because the laws of the special sciences support their instances more than any corresponding physical (disjunctive) laws, then reifying the kinds of the special sciences is rational when weighed up against scientific evidence and explanation. Importantly, Fodor construes the laws as more basic than the kinds, in his account, in so far as the kinds are parasitic on the ceteris paribus special science laws that are confirmed by their instances.

However, Fodor faces the following dilemma. On the one hand, if he construes natural kinds as more basic, then he is committed to natural kinds with essential properties and is faced with Kim's challenge of local reductionism. On the other hand, in so far as he makes the laws of the special sciences more basic, he must provide an account of why the generalisations of the special sciences can be construed as genuine cp laws[16]. As seen above, there are semantic and methodological problems in doing the latter. However, one problem that is not often considered is whether the laws of the special sciences are related to special science kinds in the way that Fodor construes them. Why given the hypothesis of MRT, do we proceed to cash out the special sciences in this way?

I wish to argue that an enquiry into the natural kinds of the special sciences particularly as they function in scientific practice reveals that they are in fact a fundamentally different kind of kind for the following reason; that every member of a special science kind has the properties it does with respect to a certain causal-historical story, which is spatio-temporally specific to a given environment or domain. It is this domain-specificity that precludes the cp generalisations that refer to the putative natural kinds of the special sciences from holding in every possible world. Millikan (1999) has called such special science kinds 'historical kinds'. She argues that special sciences kinds are defined by reference to historical relations among the members, what she terms their 'proper function'. Therefore, the functional role is determined by the historical function that has been associated with a kind in a

[16] As we saw there are both semantic and methodological problems in providing an account of putative *cp* laws of the special sciences. Fodor thinks that he can provide truth conditions for *cp* laws, which makes them both meaningful and falsifiable. See Fodor (1991) for this account.

given special science rather than in terms of special science laws. Importantly, this leaves ample room for the exceptional cases to the generalisations of special science disciplines.

Moreover, it marks a difference of kind between the natural kinds of the physical sciences and the historical kinds of the special sciences. If we accept that the kinds of the special sciences are fundamentally different from those of their physical science counterparts in terms of how their members are individuated then we see that the scientific methodology of the special sciences is quite different. Construing special sciences kinds in this way allows us to dispose of the requirement that cp generalisations ought to be construed as universally quantified laws qualified ceteris paribus.

The logical locution will be quite different. So instead of "All x's are y's cp" we will have the "The x's are y's in D". What we are saying here is that most normally functioning x's will be y's given the same causal-historical background conditions (i.e. the same domain). Importantly, since there is no universal quantifier then we can leave the possibility of exceptions open. "The x's are y's" should be understood as an inference rule, rather than a law of nature. It is an inference rule that can be tacitly used to ground narrow scope inductions. Therefore, special science kinds will be projectible within the specific domain of the special science itself. The domain of the special science will be based on a causal-historical story shared by all members of the historical kind. Consequently, observations of members of a special science kind can be used to confirm facts about other members of the same kind, be it mineralogical species in geology, biological species in evolutionary biology or whatever. When exceptions occur, it will be because the domain has changed in some way. Empirical enquiry will be able to explain how it has changed and how it has affected the historical kind.

The antecedent of the generalisation "The x's are y's in D" applies to the specific spatio-temporal domain, which applies to all members of x. Importantly however, it not an accidental generalisation, in the sense that there is a further causal-historical story as to why x's have the properties referred to by y's. There is also a further causal-historical story as to why certain x's fail to be y's. These empirical accounts constitute genuine scientific explanations as to why x possesses the properties it does when it does. This proposition has a posteriori necessity in the specific spatio-temporal domain that it refers to. This opens up the traditional dichotomy between those generalisations that are accidentally true and those that are universally true.

Moreover, this causal-historical explanation would be lost at the microphysical level. By MRT, different physical mechanisms implement the regularities referred to by generalisations that refer to historical kinds. The special science explanation is the only one that can tell the specific story required to explain how a given historical kind has the properties that it does, and also to bring together at the macro-level why such disparate physical mechanisms implement the regularities that they do.

So a new picture of special science kinds emerges from a glimpse at scientific methodology in these sciences. The generalisations of the special sciences will have narrow scope, but will nevertheless be testable within this limited domain. So for example, there will not be a special science of psychology that quantifies over every kind of psychological thing. Rather, there will be a special science of psychology that quantifies over homo sapiens. Lange (2002) has noted that in scientific practice this is how all of the special sciences work. He uses as examples the domain-specificity of enquiry in the special sciences such as "Establishment and maintenance of claw bilateral symmetry in snapping shrimps". (Lange (2002) p. 247.)

One possible problem is that under the latter construal the special sciences become descriptive rather than explanatory. It might be argued that normal science within the special sciences becomes the task of taxonomising the universe within specific domains. In other words, the task of the special sciences is a purely contingent historical one, namely that of sorting out the macro-level distribution of matter in certain spatio-temporal zones. For this reason, the special sciences become more like the social sciences than the fundamental sciences in methodology.

What is meant here is that there is a fundamental distinction between the methodologies of the social sciences and the hard physical sciences, in so far as the social sciences are idiographic, characterized by singular assertory propositions and individualising conceptualisation and the natural sciences are nomothetic, characterised by general apodictic propositions and generalising conceptualisation. (Shurz (1954) : 257). This analogy between the special and social sciences is meant to imply that the special sciences are not scientifically legitimate. The idea is that once we give up on general laws, the special sciences become idiosyncratic sciences that are specific to a human point of view. Hence, while these disciplines might be perfectly valid for describing a distinctly human world, they are not the subjects of real science, in so far as science should discover universally true explanations of the universe.

I wish to argue that this has to do with a bias about what genuine explanation involves. The idea is that an explanation that is generated from a generalisation with restricted scope (i.e. one that is domain-specific) is not legitimate. What is being assumed here is that because the kinds of the special sciences are only projectible within a narrow scope that they are cannot appear in genuine nomological explanations. The inference then is that since the special sciences (so-described) do not generate nomological explanations, then they are not scientifically legitimate.

Cp Laws and Explanation

This brings me to the chief assumption about legitimate science, namely the assumption that genuine scientific explanations must involve laws of some sort. What is being assumed here is that all explanation should involve either induction or deduction from universal or statistical laws and initial conditions, namely; a covering-law model of explanation. Clearly then another motivation for the unrelenting campaign for ceteris paribus laws that quantify over real natural kinds is an attempt to fit the explanations of the special sciences into some sort of covering law model of explanation[17].

So what is this assumed ceteris paribus covering law model? The aim is to illustrate how special sciences explanations can be deduced from particular initial conditions in conjunction with genuine ceteris paribus laws. The Deductive-Ceteris Paribus model can be formulated as follows:

D-CP CP_1 CP_2 CP_3 Ceteris Paribus Laws

C_1 C_2 C_3 + Relevant initial conditions (Explanans)

F_1 Fact to be explained (Explanandum)

According to this model, the fact-to-be-explained can be straightforwardly deduced from the ceteris paribus laws in conjunction with relevant initial conditions.

Let's consider an example from a special science like Epidemiology:

"All epidemics increase their incidence rates exponentially ceteris paribus" (Farr's Law)

X is an epidemic

X increases exponentially.

As seen earlier there is a semantic problem with postulating laws to corroborate special science explanations in that exceptions can occur that contravene the law. The law can yield true explanations given some initial conditions and false explanations given others. Moreover, it may even yield true explanations some of the time and false explanations other times. Consider the following examples:

---

[17] For a detailed version of this argument see Tobin. E. (forthcoming) 'Towards a New Model Of Explanation for the Special Sciences', *Causality & Explanation Essays in Honor of Wesley Salmon,* Autonomous University Press.

"All Epidemics increase their incidence rates exponentially ceteris paribus" (Farr's Law)

Bubonic plague is caused by the germ Yersinia Pestis

Bubonic Plague will increase exponentially.

"All Epidemics increase their incidence rates exponentially ceteris paribus" (Farr's Law)

Polio is a viral infection

Polio increases exponentially.

In the first example, we can imagine cases in which the explanandum is true such as the Black Death in the 14th Century. However, we can also imagine how a change in environmental factors may make it false in other situations. For example, the bubonic plague may not increase exponentially if certain preventative measures are taken. For example, health authorities can ensure that all people who are exposed to plague are given antibiotic treatment for seven days, or if hygiene and environmental sanitation are extremely good then the chances of epidemic are rare. So, given this particular set of initial conditions the law is only true "in the absence of disturbing factors".

The second example is a case in point where an epidemic has not increased exponentially, in other words where the explanandum is false. Since, Polio has been ubiquitous since the beginning of human history, its incidence rate should have peaked a long time ago. However, improvements in sanitation have meant that outbreaks of Polio have occurred most rapidly in the last century. In highly unsanitary circumstances, most children are exposed to the virus during infancy, when infection with polio is asymptomatic, and this results in lifelong immunity. For, this reason, increases in sanitation have meant that people not immuned to the disease contract it later in life and become infected by it when it is more likely to be symptomatic.

For this reason, it can be argued that whether or not a given ceteris paribus law will hold is highly context-specific. We cannot say what conditions must hold in order for all "all other things to be equal" nor can we say what conditions must intervene to make all other things not equal". It can be argued that we do not know what it means when we say that a law holds ceteris paribus. This is to argue that ceteris paribus laws are vacuous, in so far as they have no empirical content.[18] We can only accommodate exceptions on a case-by case basis.

[18] This is the standard semantic objection to the construal of *ceteris paribus* laws. See Earman, J. & Roberts J. (1999) '*Ceteris Paribus* There Are No Provisos,' *Synthese*

Since, the special science ceteris paribus clause cannot be 'theoretically' completed, then any deductive explanation containing cp laws is going to contain an unstatable first premise. Clearly, according to this model, if you cannot articulate the premises of the argument, then you don't generate an explanation based on those premises. Therefore, the deductive ceteris paribus model does not validate special science explanations in so far as we understand the generalisations of the special sciences as vague and imprecisely formulated.[19]

Clearly, then a ceteris paribus covering law model is not the correct model of explanation for the special sciences. Importantly, if we reject the covering-law model, then the denial of ceteris paribus laws need not lead to the demise of the special sciences. In fact, only then will we make progress on the real issue, namely that of giving an account of the explanations of the special sciences even though they are not law-like. It is clear that a new model of explanation is needed.

A glimpse at the role of historical kinds in domain specific generalisations provides the starting point towards the formulation of such a model. Take any inference rule in the special sciences 'The F's are G's in D'. Take Farr's law in epidemiology again, which states that epidemics must increase their incidence rates exponentially. This will be re-phrased as "Epidemics increase

---

118 438-479. Earman, J. Roberts, J. T. & Smith, S. (2002) '*Ceteris Paribus* Lost', *Erkenntnis,* Vol. 57, No. 3.

[19] It is important to point out that this paper is not arguing against explanations generated by *ceteris paribus* laws *tout court*. Rather, if there were genuine *ceteris paribus* laws then it would be possible to determine what the "all other things being equal" refers to. In this respect, we would not have an unstable premise and it would be possible to generate explanations by means of the deductive *ceteris paribus* model. Therefore, there may be genuine *ceteris paribus* laws in physics for example, if we could complete the *explanans* such that we can say what background conditions would have to be in place in order for the *explanandum* to occur. This is clearly Cartwright's (2002) account of *ceteris paribus* laws she states "The laws I talk about either can be stated in precise and closed form or they entail strict or statistical regularities in the course of events or both" (Cartwright (2002) : 149.) Clearly, then the deductive-*ceteris paribus* model can either be formulated in a similar way to the D-N model or to the D-S model. According to Hempel's D-S model, statistical laws can be used deductively in so far as they assume further strict laws. In a similar vein, Cartwright's analysis suggests that *ceteris paribus* laws can be used to generate explanations either because we can complete them or because they depend upon further statistical laws. Importantly, for the purposes of this paper, I contend that the putative *ceteris paribus* laws of the special sciences do not fit into either of the deductive-*ceteris paribus* approaches, since they do cannot be stated in closed form and they do not entail statistical regularities (i.e. we have no idea of the probability with which they hold).

exponentially in domains that favour the spread of the virus or bacterium". The point is that the domain will be set by the initial conditions at time t_1. So, if the initial conditions are the Polio Virus as it pertains to First World countries in the 20th century then we can explain why the epidemic will fail to increase exponentially.

Therefore, the generalisation is relative to the proviso of normal conditions. What we are seeing here is that normal conditions can and do change. This domain specificity of the special sciences explains, why the historical kinds have vague boundaries and why the special science have no universally quantified laws. Normal scientific practice in the special sciences proceeds by explaining why most instances of a given generalisation are projectible within a given domain. Also more importantly, explanations in the special science do something that covering law explanations cannot do, namely, they explain why some instances fail to satisfy the regularity picked out by the generalisation.

Some special science generalisations will have a greater degree of invariance under change of relevant factors than others. Assessing the degree of invariance under change within scientific practice is what makes the special sciences special[20]. Once a generalisation is invariant across a given domain then we can use it to make inductive inferences within that domain. However, the possibility of exception cases is always left open. The task of empirical enquiry is to assess what changes the generalisation is not invariant under in scientific practice and why. This is how normal scientific practice makes progress in the special sciences. Hence, generalisations are corroborated neither by cp laws that define the causal role of members of a natural kind, nor by their role in covering-law explanations. Once we reject the former possibilities the ground is cleared towards a better understanding of scientific methodology in the special sciences.

Clearly, then MRT only provides a starting point for assessing the special/physical science distinction. Once, we see that the special sciences work in rather different ways to the physical sciences, it is clear that the move from no special sciences in the strictest sense to no special sciences tout court is an illegitimate one to make. Once, we assess the methodologies of the special sciences, we see that the disunity of the sciences is also marked by a difference in kind. The special sciences unify causal, historical and contextual explanations of the macro-world. The special sciences so described not only describe true facts about the macro-world, they also

---

[20] For a detailed account of invariant *cp* generalisations see Woodward, J. (2003), *Making Things Happen: A Theory of Causal Explanation*, Oxford, Oxford University Press.

explain why they came about historically causally and contextually i.e. within their own restricted frame of reference.

So to begin answering the question posed by the paper title, what makes a special science special? We can see that it is not only its irreducibility to the fundamental sciences, but also their difference in scientific method, in so far as the explanations that they provide work in an altogether different way to the explanations of the fundamental sciences. The special sciences require generalisations (not universally quantified) that show why a historical kind is characteristic of a given kind within a certain domain. In explaining why exception cases occur, the special sciences can classify the range of invariance that each generalisation has. Importantly, explanations in the special sciences tell us something about the macro-world that is lost at the level of fundamental physics. Crucially, because the explanations of the physical sciences and the special sciences are of a different kind, they can be complimentary rather than mutually exclusive.

ACKNOWLEDGEMENTS

Thanks to Dr. Richard Gray and Prof. William Lyons for their comments and suggestions on earlier drafts of this paper. Thanks are also due to the audience at Queens University Belfast in April 04' for their interesting questions and suggestions.

Bibliography

Cartwright, N. (1983). How the Laws of Physics Lie, Oxford: Clarendon Press.

———. (2002), 'In Favor of Laws that Are Not Ceteris Paribus After All', Erkenntnis, Vol. 57, No. 3: 425-439.

Earman, J. Roberts, J. T. & Smith, S. (2002), "Ceteris Paribus Lost", Erkenntnis, Vol. 57, No. 3: 281-301.

Earman J, & Roberts (1999), 'Ceteris Paribus, There are No Provisos' Synthese, 118: 439-478.

Fodor, J.A. (1974), 'Special Sciences or the Disunity of Science as a Working Hypothesis', Synthese 28:97-115 or Reprinted in Fodor J.A. (1981) Representations, Cambridge: M.A. MIT Press.

———. (Jan, 1991), 'You Can Fool Some People All of The Time, Everything Else Being Equal; Hedged Laws and Psychological Explanations', Mind, 100, 1 : 19-34.

———. (1998), 'Special Sciences: Still Autonomous after All these Years', In Critical Condition: Polemical Essays on Cognitive Science and the Philosophy of Mind, Cambridge, M.A. MIT Press : 9-24.

Hacking, I (1991a), 'A Tradition of Natural Kinds', Philosophical Studies, 61 : 109-126

Hempel, C. G. (1965), Aspects of Scientific Explanation, New York: Macmillan.

Lange, M. (2000), Natural Laws in Scientific Practice, Oxford: Oxford University Press.

Kim, J. (1992). 'Multiple Realisations and the Metaphysics of Reduction, Philosophy and Phenomenological Research, Vol. 52, No. 1 : 1-26.

Millikan,R. (1999), 'Historical Kinds and the Special Sciences' Philosophical Studies, 95 : 45-65.

Rosenberg, A. (1985) The Structure of Biological Science. Cambridge: Cambridge University Press.

Shapiro, L, (Dec. 2000), Multiple Realizations', Journal of Philosophy, Vol. XCVII, No. 12: 635-654.

Schiffer, Stephen (1991), Ceteris Paribus Laws, Mind 91: 1-17.

Schurz, A. (1954), 'Concept and Theory Formation in the Social Sciences', The Journal of Philosophy : 257-273.

Stern, C., and E. R. Sherwood. (1966), The Origin of Genetics: A Mendel Source Book. New York: W. H. Freeman.

Woodward, J. (2003), Making Things Happen: A Theory of Causal Explanation, Oxford: Oxford University Press.

CHAPTER TWO: EOIN RYAN

THE EVOLUTIONARY STANCE

Daniel Dennett is famous for his 'intentional stance' in philosophy of mind. His idea of stance is an outgrowth of a naturalizing project, a reconciliation (rather than reduction) of the languages of folk psychology/intentionality and natural science. This was originally outlined in Content and Consciousness (1969/1986), and in that book Dennett points to the importance of evolutionary theory for his project. Although this importance seemed to be downplayed for a time, in much of his recent work evolution has loomed very large. Dennett actually proposes three stances, viz the physical, design, and intentional stances, but his presentation of these often seems to imply that the physical stance is the 'realest' one, and that there are only practical reasons for our not seeing the reductions from the other stances to the physical level. In fact, practicality is important to Dennett, but he isn't just being glibly instrumentalist. Sometimes the view revealed from the physical stance may seem the realest position, but sometimes Dennett explicitly and strongly asserts the reality of the other stances. In this paper I argue that Dennett's is an ontological position. We normally take physical description to be the basic description of what there is in the world, but Dennett makes evolution the joint arbiter of 'reality', with respect to the reality of mental states (his original project) and on a more general scale.

In Content and Consciousness Dennett hopes the theory of evolution by natural selection can give substance to an account of intentional content (p.41). Unfortunately, he cannot find a "neat synonymy or correlation between Intentional and non-Intentional sentences" (p.89). However, he can give some "sense [to] the lesser claim that certain types of physical entities are systems such that their operations are naturally to be described in the Intentional mode".

Dennett therefore will rely on understanding what we mean by 'natural' description or explanation. This might hint at a conceptual-scheme metaphysics, but Dennett is leery of capital-M Metaphysics. He gives a lot of space to dismissing the ontological battles in philosophy of mind, which are typically stand-offs between physicalism and dualism. Such controversies stem from category mistakes. For instance, a physicalist, unable to identify the mental with anything physical, decides the mental is nothing at all; a dualist, unable to identify the mental with anything physical, decides the mental is something else entirely. Both are tied into exactly the same unhelpful knot (pp.6ff). We should realize instead that natural explanation need not worry about ontology in this way; it just isn't helpful.

What is helpful are different stances in different situations. It is important that taking a stance does not involve ontological commitment: the stances are simply routes to useful explanations and predictions, adopted on purely pragmatic grounds (1987, p.15). As soon as a stance ceases to be useful we should try another. Thus, Dennett's idea of natural explanation is linked to usefulness in the natural world

Dennett suggests that our working explanations of the world are given in terms of one or other among the physical, design, and intentional stances. If we can comprehend all relevant physical variables we can use the physical stance. However, although we could conceivably explain what a kidney, say, is doing in terms of the minutiae of physical events that constitute it, it is usually more convenient to talk in terms of the function for which it is 'designed'. Further, for many complex systems perspicuous explanations are best given by regarding them as agents, that is by attributing intentional properties such as beliefs and goals to them.

These stances are to be taken as only attributions. The intentional stance does not need to claim its attributions betoken real mental entities. The design stance need only say what the proper function of a system seems to be. And the physical stance—well, what of the physical stance?

The physical stance is also, nominally at least, only a stance, but Dennett never seems to make much of this. On the contrary, although his stances may entail no ontological commitments, they do seem to fall into a hierarchy in which the physical stance seems the most basic. Resolute pragmatism requires that once a stance becomes ineffective we use another. When the intentional stance fails, we try explanation in terms of design. When the design stance fails, we try the physical stance.

When pragmatism commands that we change stance, we always change in one direction. When a machine (or a biological organ) fails, the design stance is by definition irrelevant, because the system is no longer functioning as it is meant to. We must change stance, so we look to physical explanations. We certainly do not explain malfunction in terms of the intentional stance because—Murphy's Law notwithstanding—most gizmos don't stop working out of sheer malice.

All this would seem to suggest that both the intentional and design stances are just useful façons de parler: summarizing explanations that conceal the laborious details of the real physical explanation.

Sometimes Dennett seems to admit as much. No intentional or functional entity is ever perfect; things often go wrong for it. Therefore, if we rely on the intentional and design stances we will stumble whenever things do not go as they should. By contrast, if we could only make a go of physical explanation, it would always work. Dennett has said that a

> difference between the physical stance and the design stance is that no oxygen atom is 'defective' ... They are all 'perfect'. If they weren't, there would have to be a physical design stance of sorts (2000, p.349).

'Perfect' oxygen atoms will always do what they do, so physical predictions of what they do can also be perfect in principle. The rather clumsy distinction between the 'physical stance' and a putative 'physical design stance' seems to admit that the physical stance is not merely a stance, or else, that the other stances are merely stances.

Need this be an expensive admission on Dennett's part? After all, he has provided an explanation of why we use different levels of explanation (practicality) and pointed out that we shift levels when the level we are at is no longer useful. Nevertheless, if Dennett's naturalism turns out to be an elaborate physicalism, then this may jibe with his declared antipathy to ontology and, more importantly, we could wonder why we should make such a fuss about the stances.

The reason we should make such a fuss is because the different stances pick out real things. Dennett invites us to imagine some super-physicists who are sufficiently brilliant to meet Laplace's challenge of predicting the events of the world from its physical state description[21], and sufficiently old-skool to be Martians. Ex hypothesi these Martians predict everything that happens (1987, p.23). Nonetheless,

> if they did not also see us as intentional systems, they would be missing something perfectly objective: the patterns in human behaviour that are describable from the intentional stance, and only from that stance, and that support generalization and predictions (p.25).

This is a puzzling. Here Dennett retains the notion that the physical stance (in which the Martians are expert) is fundamental, for the Martians predict everything that happens. But the intentional stance seems also to be fundamental because (and surely this is rather bizarre) it picks out things that are not included in the 'everything' witnessed by the Martians, for what else could 'perfectly objective' patterns that 'support generalization and

---

[21] Laplace suggested that an "intelligence that knew, for a given instant, all the forces acting in nature, as well as the positions of all the things that constitute it, and who was capable of subjecting these data to analysis, would embrace in a single formula the motions of the largest bodies and those of the lightest atom" (quoted in (Torettii, 1999, p.94), from Laplace in *Œuvres complètes*, vol. VIII, pp.vi-vii, 1795). For such an intelligence, "nothing would be uncertain".

predictions' be but real elements of a comprehensive explanation of the world.

This returns Dennett to concerns about natural explanation, and perhaps to a situation where our conceptual-scheme makes the world in its own image, for the Martian physicalists see one world, we Terran intentionalists another. Of course, what the Martians miss are patterns in human behaviour, but these patterns are themselves visible as quite large-scale physical processes including, in the particular story Dennett tells, the movements of several motorcars and people over relatively significant distances. It is our ability to attribute intentionality to human minds that allows us predict behavioural patterns resulting from such intentionality. Without this knack, such patterns seem to be invisible, and at least without significance.

Hence, a key word in this story seems to be 'pattern'. The Martians see the movements of basic physical constituents, but they miss the patterns that are formed with these elements. In his paper "Real Patterns" (1998), Dennett uses a type of artificial universe built using computer simulations of 'cellular automata' to look at the reality of patterns.

This universe is a square grid of cells. Each cell can be 'on' or 'off'. Time moves forward in this universe in discrete steps. At each time step the state of each cell is re-evaluated using a very simple rule such that the state of each cell at time t_n depends on a count of the 'on' cells among its eight immediate neighbours at time t_{n-1} (such that different counts apply different variants of the rule, for example, setting the cell to 'on', or setting it the same/opposite of what it was)

For this universe we can emulate the position of the Martian super-scientist, for we can calculate the state of each cell at any time (this is exactly what the computer simulation does). However, we can also observe that interesting patterns formed from groups of cells sometimes arise. Then we may be able to use an understanding of pattern-level behaviour to predict future states of some cells without having to go through the Laplacean calculation. For instance, if a pattern of 'on' cells which appears to move across the grid appears (a pattern that is called a 'glider'), we can easily extrapolate its future path, which is a way of predicting the future states of cells along that path.

Does it make sense to say these patterns are real? Dennett says in 'Real Patterns' that he is a mild or intermediate realist with regard to them. Observing the patterns is helpful for prediction without recourse to calculating the full state-description, but the pattern is also only a happenstance configuration of cells whose 'on'/'off' condition can be worked out more completely the Laplacean way.

However, Dennett does claim that these patterns are real (and in a more than epiphenomenal sense). It may be possible to make sense of this by considering that the universe the Laplacean describes is an extensional one.

The Laplacean intelligence knows "all the forces acting in nature, as well as the positions of all the things that constitute". The example of the grid universe in 'Real Patterns' makes this very clear. Every cell is exactly known, and the rules by which the cells change state are also known. But this is corresponds to a viewpoint outside the universe, a view from nowhere.

A view from such a position is obviously impossible, so if metaphysics is based on a view from nowhere, then Dennett's refusal of ontology makes sense. But he will claim to be a 'mild realist' and maybe this translates as a type of mild metaphysics, an attempt to do ontology from within the universe. Concomitantly, an extensional physics in its fullest sense is unavailable. So all descriptions of the universe are intensional.

The problem with this is that intensionality and intentionality are related to the imposition of concepts, and this appears to be a non-naturalist position. With regard to this, the explanations available from evolution are useful[22]. Evolution provides a mechanism for using evaluative criterion in a natural way. Norms appear in retrospect. Successful entities are those entities that survive. In this sense, patterns become real for they are stable existents within a universe and their actions are the actions that grant them this stability.

Does it make sense to ask about the ontological status of these surface patterns? Dennett, in the spirit of his earlier rejection of ontology, says, no, not really. The main thing is that observing the patterns is helpful for prediction without recourse to computing a complete state-description. We can see from the position and trajectory of a glider what the value of cells along that trajectory will be for the next several time steps. This saves us a lot of trouble, but such "intermediate realism" (1998, p.120) hardly seems satisfactory if we remember that the glider is only a happenstance configuration of cells whose real ontology is described in terms of 'on'/'off' bits governed by rules that make no mention of any coincidental surface pattern. It might be clever of us to get a qualitative feel for what's going to happen next by looking at these surface patterns, but super-Laplacean Martians (or our own computer simulation of Life) can compute exactly what will happen.

It may be possible to push a stronger metaphysics on Dennett. We must not forget that the artificiality of the Life universe means we can stand outside it with a view from nowhere. No entity within the universe has that

---

[22] 'Evolution' here is not read as the detailed theory of biological evolution on this planet. Rather it is the basic strategy exploited by that theory, a strategy which allows explanation of change and development without reference to anything other than reproduction with the possibility of random mutation in a competitive environment, which is both quite a simple and a distinctly purposeless algorithm.

point of view. And if we consider the nature of our view from nowhere, we might doubt its validity. The Laplacean stance enabled by this external viewpoint relies on vast computation to predict what happens next. In fact, the Laplacean stance does not predict what happens next, it simply states what happens next. If our Laplacean computer is the computer on which the Life simulation is running, then this statement is what happens next! (There is no greater computational machine to shortcut this process and make faster predictions.) Explanations of what happens in this world in terms of complex entities (rather than in terms of the overall array of cells), whether posited from outside or inside the universe will invoke notions of causation and explanation. By contrast, the Laplacean calculation does not need to mention causation. Its purely extensional view simple lists a progression of states.

A pattern that survives in the competitive context of this universe is a configuration of cells that happens to persist. But once it has persisted we can suggest why that configuration rather than another survived. For example, if a pattern is protected from an assault of other, 'eater', patterns by its thick walls, neither it, its walls, nor the eaters actually exist from the point of view of the physics of the universe. Yet patterns with thick walls will survive more often that those without. This level of explanation holds from perspectives internal and external to the universe, and it is the level of evolutionary explanation.

Dennett rejected ontological questions. They were asked in the wrong terms, he said, and instead his stances focus on intelligible natural explanation. But intelligible explanation seems to entail notions of ontology, for it involves entities that do things. However, it is important also that they do things for a reason. A truly extensional picture of the universe might well drop notions like causality, for physics would simply unfold. However, this is not incompatible with evolutionary accounts of how and why entities or patterns persist and develop on the surface of this physical structure. The idea of actual extensional explanation does not quite make sense, evolution shows how explanation, which is a matter of stances, can approach extensional explanation in the limit. So, so far as 'ontology' is to make sense, it should be construed first in terms of the surface entities and only then in terms of the physical structure, making evolution and physics equal partners in the generation of the types of entities open to explanation.

Bibliography

Dennett, D. 1978. "Intentional Systems" Brainstorms: Philosophical Essays on Mind and Psychology. Hassocks, Sussex: Harvester Press. Pp.3-22

———. 1986. Content and Consciousness. 2nd edn. London: Routledge. (First published 1969.)

———. 1987. "True Believers" The Intentional Stance. Cambridge, MA: MIT Press. Pp.13-35.

———. 1996. Darwin's Dangerous Idea: Evolution and the Meanings of Life. London: Penguin.

———. 1998. "Real Patterns". Brainchildren: Essays on Designing Minds. London: Penguin. Pp. 95-120.

———. 2000. "With a Little Help from my Friends". Dennett's Philosophy: A Comprehensive Assessment. Don Ross, Andrew Brook, David Thompson, eds. Cambridge, MA: MIT Press. Pp. 327-388.

Torretti, Roberto. 1999. The Philosophy of Physics. Cambridge: Cambridge University Press.

CHAPTER THREE: STEFANO GATTEI

THOMAS KUHN AND THE LEGACY OF LOGICAL POSITIVISM

One man come he to justify
One man to overthrow
Paul Hewson

Orthodoxy, especially in light of the enormous impact of Thomas Kuhn's ideas, presents us with a picture of a sharp and massive break, of a true revolution: Kuhn is viewed as a philosopher whose main achievement is to have undermined a whole philosophical tradition. I think this is wrong. From many and often fundamental points of view Kuhn did not manage to break entirely with the preceding philosophical tradition of Logical Positivism (or Logical Empiricism): his works are laden with principles belonging to that very empirical philosophy he was determined to reject. Furthermore, only a partial challenge of positivism and empiricism can actually account for the genesis of Kuhn's philosophical perspective – incommensurability, the notion of progress, the rejection of the concepts of truth and verisimilitude, and the very thesis of "world change" (one of the theses deemed most radical and characteristic of Kuhn's philosophical stance) are all consequences of the empiricist elements that his philosophy retains. Appearances to the contrary notwithstanding, the implicit presuppositions and the stated principles of Kuhn's philosophy are not very different from those of logical positivists or logical empiricists he saw himself to be distancing from.

Far from spelling the "demise"[23] of Logical Positivism, then, Kuhn's philosophy is its natural continuation (which is, perhaps, the reason why Carnap welcomed his book so favourably[24]). Like Copernicus, who, while fetching the first fatal blow to the Aristotelian-Ptolemaic worldview, was also

---

[23] See Friedman (1991), p. 1; for a criticism, see Gattei (2000) and (2001), pp. 818-822.

[24] See Carnap's letters to Kuhn, published in Reisch (1991); for the relationship between Carnap and Kuhn, see also Irzik, Grünberg (1995), Friedman (1991), (1992), (1993) and (2001), pp. 18-19, 22, 41-43 and 56, Earman (1993), Axtell (1993), Irzik (2002) and Gattei (2004), chapter 5. In a recent article, Michael Friedman explicitly outlines his project, namely, to rescue Logical Positivism through a reassessment of Kuhn's philosophy: see his (2003),

irrevocably soaked in that very same way of thinking, so Kuhn can be regarded as the last exponent of the philosophical tradition he was determined to reject. In The Copernican Revolution Kuhn wrote that Copernicus is like the middle point of a bend in a road: from one point of view it is the last point of one stretch of the road; from the other, it is the beginning of the next. In this sense, Kuhn is the last of the neopositivists and, at the same time, the first of their successors. In particular, Kuhn's later "linguistic turn" marks a clear a step back into inductivism and justificationism, completely disregarding Popper's philosophical revolution and turning Kuhn's followers into foundationalist philosophers.

Having dealt with the complex relationship between Kuhn and Logical Positivism more broadly elsewhere,[25] in the present paper I shall focus on what I take to be a central issue to understand both such relationship and the Popper-Kuhn debate. It is the notion of truth.

Kuhn on truth and consensus

In the first edition of The Structure of Scientific Revolutions (1962) Kuhn hardly refers to the concept of truth: he has no need of it, not even in order to characterize and explain progress.[26] In the 1969 "Postscript" to the second edition of the book he introduces two arguments against the notion of truth

---

[25] In my Gattei (2004).

[26] See Kuhn (1962), pp. 170-173. In fact, in the first edition of *The Structure of Scientific Revolutions* the term "truth" appears only in a quotation from Francis Bacon (on p. 18). In the closing pages of the book, Kuhn writes: "The developmental process describe in this essay has been a process of evolution *from* primitive beginnings – a process whose successive stages are characterized by an increasingly detailed and refined understanding of nature. But nothing that has been or will be said makes it a process of evolution *toward* anything" (Kuhn (1962), pp. 170-171). He then urges us to give up the concept itself in order to get rid of some of the problems which have afflicted the history of Western thought: "We are all deeply accustomed to seeing science as the one enterprise that draws constantly nearer to some goal set by nature in advance. But need there be such goal? Can we not account for both science's existence and its success in terms of evolution from the community's state of knowledge at any given time? Does it really help to imagine that there is some one full, objective true account of nature and that the proper measure of scientific achievement is the extent to which it brings us closer to that ultimate goal? If we can learn to substitute evolution-from-what-we-know for evolution-toward-what-we-wish-to-know, a number of vexing problems may vanish in the process. Somewhere in this maze, for example, must lie the problem of induction" (*ibid.*, p. 171).

implicit in the traditional view of progress as increasing verisimilitude.[27] Later, he relates his rejection of truth to incommensurability.[28]

Indeed, as a consequence of Kuhn's later characterization of incommensurability, which is attributed both an inevitable and a functional role for the growth of scientific knowledge, there is no need for the notions of "truth" and "approximation to the truth". Kuhn has always opposed the correspondence theory of truth and criticized its applications to the relation between scientific theories and reality: as history can show, he says, there is "no theory-independent way to reconstruct phrases like 'really there'; the notion of a match between the ontology of a theory and its 'real' counterpart in nature [...] seems to me illusive in principle. Besides, as a historian, I am impressed with the implausibility of the view. I do not doubt, for example, that Newton's mechanics improves on Aristotle's and Einstein's improves on Newton's as instruments for puzzle-solving. But I can see in their succession no coherent direction of ontological development".[29] In its stead, he argues, a weaker concept of truth can be retained: we can talk about truth only within the context of a given lexicon.[30] Indeed, within such a lexical structure a claim may be properly said to be true or false, but that does not hold for the system of categories embedded in the lexicon, which is not itself capable of being true or false.[31] Truth is internal to lexicon in the sense that its use is

---

[27] See Kuhn (1970a), pp. 205-207. "A scientific theory is usually felt to be better than its predecessors not only in the sense that it is a better instrument for discovering and solving puzzles but also because it is somehow a better representation of what nature is really like. One often hears that successive theories grow ever closer to, or approximate more and more closely to, the truth. Apparently generalizations like that refer not to the puzzle-solutions and the concrete predictions derived from a theory but rather to its ontology, to the match, that is, between the entities with which the theory populates nature and what is 'really there'. Perhaps there is some other way of salvaging the notion of 'truth' for application to whole theories, but this one will not do" (Kuhn (1970a), p. 206).

[28] See Kuhn (1989), (1991), (1992) and (1993).

[29] Kuhn (1970a), p. 206, emphasis added. Such expressions like "zeroing in on the truth", or "getting closer and closer to the truth" are meaningless as a consequence of incommensurability: see also Kuhn (1993), p. 330.

[30] The notion of truth involved may be provided, Kuhn suggests, by "something like a redundancy theory of truth" (Kuhn (1991), p. 8).

[31] Reaffirming his opposition to the traditional, correspondence theory of truth, Kuhn sketches his position about the nature of truth in the context of the idea of variant lexical structures. Scientific theories are the source of alternative sets of taxonomic categories which are imposed on the world by theories. Such categories do not reflect reality: it is impossible for them to do so – they are, at most, ways of ordering experience (taxonomy). The sets imposed on the world depend on theories and vary with them. In a Kantian way (but with mobile categories), the taxonomic categories of the scheme provide a structure for possible experience. Thus we can say, following

restricted to assessing claims made within the context of the lexicon: truth claims in one lexicon are not relevant for those made in another, nor truth can be applied to a lexicon itself. In other words, Kuhn rejects the idea that the structure which constitutes the theory might reflect the way the world is, independently from theory. The lexicon has the status of a linguistic convention which may be judged only on the basis of how well it serves a particular purpose, rather than how well it reflects reality. As a result, though Kuhn assumes the existence of an independent reality throughout his work, his position involves idealistic leanings.[32]

"Lexicons are not [...] the sorts of things that can be true or false", writes Kuhn:[33] their logical status is that of general meanings, that of a convention which we can justify only in a pragmatic way. A lexicon marks the distance between the reality as described by a theory and the theory which describes it in different ways: "Experience and description are possible only with the described and describer separated, and the lexical structure which marks the separation can do so in different ways, each resulting in a different, though never wholly different, form of life. Some ways are better suited to some purposes, some to others. But none is to be accepted as true or rejected as false; none gives privileged access to a real, as against an invented, world. The ways of being-in-the-world which a lexicon provides are not candidate

---

Kuhn's own image, that as theories change, the world changes with them. See Kuhn (1991) and Sankey (1997), chapter 3.

[32] See Hoyningen-Huene (1989a, 1993), pp. 267-271, and (1989b); see also Sankey (1997), chapters 2-4.

[33] Kuhn (1993), p. 330. See Wittgenstein (1969), §205: "If the true is what is grounded, then the round is not *true*, nor yet false"; and "[...] why should the language-game rest on some kind of knowledge?" (§477; see also §559). According to Wittgenstein, a language game presents no gaps, since together with its possible moves it also defines the space which makes those very moves possible: just as the rules of the game define which moves belong to it, so the grammar of the language circumscribes what is meaningful. Nothing meaningful can therefore remain outside its boundaries and establish itself as a mark of the incompleteness of the language game (*incommensurability*). A game to which new rules are added is not a richer game, but simply a new game (*paradigm shift*). Therefore, a language game is criterion to itself – like the sample standard metre preserved at The International Bureau of Weights and Measures of Sèvres, near Paris, it is not itself measurable, since it is not possible to measure what is to be the unit of measurement: its having a length can not be ascertained, but it is a feature which displays itself in the way we use it when measuring (see Wittgenstein (1953), part I, §50, and §241). "If you tried to doubt everything you would not get as far as doubting anything. The game of doubting itself presupposes certainty" (Wittgenstein (1969), §115; see also §160, §450 and §625. Doubting the paradigm means, on the one hand, condemning oneself to silence; on the other, it means extending the practice of doubt beyond what is reasonable (*i.e.* meaningful) to doubt.

for true/false":[34] they are assessed on the basis of their ability to serve a particular function, not to reflect reality.

Just as Kuhn, following Wittgenstein, can understand mistakes only in terms of established rules,[35] so can he understand truth only within the context of a given linguistic framework. This is one of Kuhn's major holdovers from positivism. Positivists also placed particular emphasis on community: they regarded communal collaboration as important for the production and justification of scientific knowledge, which they in turn regarded as important for the unity of science. It is this very emphasis that fuels Kuhn's conception of science as a social institution, and his attempt to define scientific knowledge, if not truth itself, in terms of the consensus of belief that is forged among its members.

The strength of a community and the solidarity of belief sounds attractive because it offers ease and disengagement. But it is a two-edged sword, for when we add to Kuhn's ideas that scientists must commit themselves uncritically to a paradigm and that it is appropriate for the profession to ostracize those who disagree, then it becomes difficult to see how science can uphold freedom of thought.

The poverty of institutionalism

Kuhn's position is rooted both in justificationism and in a particular way of posing problems which is typical of Wittgenstein and his followers.[36] Taken together, these two closely interwoven aspects work together and reinforce one another, forcing the compartmentalization of knowledge and the limitation of rationality.

From David Hume (1711-1776) onwards, it has been asserted that there are two kinds of inference: deductive inference, which defines logic; and

---

[34] Kuhn (1991), p. 12.

[35] That is why, for example, Kuhn chooses the term "puzzle" instead of "problem" (akin to Wittgenstein's choice of "game"): in such a view, no room is left for a critical form of rationality that is independent from, let alone opposed to, the dominant epistemic practices of the time. A crucial point that unites Kuhn and logical empiricists is their unwillingness to associate radical criticism, or a choice between alternative normative systems, with anything they would call "rational": for them, rationality is to be found not in making or breaking a rule, but merely (*à la* Wittgenstein) in *following* a rule. The difference between making and following a rule corresponds to the descriptive and the prescriptive sides of the normative enterprise.

[36] Following William Bartley's suggestion, we could call the latter "Wittgensteinian problematic": see Bartley (1990), chapters 14 and 15.

inductive inference, which defines the natural sciences.[37] The two apply, so to say, to different fields, and must not be confused: the problem of induction is simply dissolved once we learn not to apply the standards of deductive logic to judge inductive inference. Once we realize that the two principles cannot be unified, the task of the philosopher is simply that of describing and clarifying the standards of deductive and of inductive reasoning. Most positivists, while maintaining the unity of the sciences, accepted this "methodological" division. Wittgenstein extended this approach: each discipline, or field, or "language game", or "form of life" is alleged to have its own standards, or principles, or "logic", which need not conform to or be reducible to any other standards and which, again, is the special task of the philosopher to describe and clarify – not the least to judge, defend or criticize.[38] There is no arguing or judging among disciplines: criticism, evaluation and explanation would no longer be proper philosophical aims. Knowledge is essentially divided, and description is all that remains to the philosopher. All he can do is to describe the logics, grammars or first principles of the various kinds of discourse, and the many sorts of language games and forms of life in which they are embedded. Philosophical critique is no longer of content, but of criteria application: as Paul Feyerabend put it, all that is left are "consolations for the specialists".[39]

---

[37] "Instead of being a faulty sort of deduction, induction is fundamental, defining science – just as deduction is fundamental, defining logic" (Bartley (1990), p. 219).

[38] See Wittgenstein (1953), part I, §66: "don't think, but look!"; "[…] we may not advance any kind of theory. There must not be anything hypothetical in our considerations. We must do away with all *explanation*, and description alone must take its place" (§109); see also *ibid.*, §89, and his (1969), §189. It is no accident that Wittgenstein had already used the term "paradigm" to refer to what rules in an activity very similar to Kuhnian "normal science": see his (1953), part I, §50 and §54.

[39] See Feyerabend (1970). "Consolations for the Specialist" is the title of Feyerabend's critical remarks on Kuhn's paper, as delivered at the 1965 Bedford Colloquium in London (see Kuhn (1970b); a revised and expanded version of Feyerabend's paper is his (1978). The affinities between Kuhn's position and Wittgenstein's are substantial: see, in particular, Wittgenstein (1921), 4.112; (1953), part I, §109; and (1969), §189. To Wittgenstein, philosophy has no cognitive function – rather, it has a "therapeutic" function (see his (1953), part I, §109, §133 and §255). The descriptive task which characterizes philosophy concerns the rules governing the use of our language, that is, the grammar of the terms that constitute it: "description" refers to the description of language games, and it aims at showing the rules of those games and hence the structures which characterize them. Concerning rules, and not facts, description has an exemplary value. There is a close parallelism between the role of Kuhn's "exemplars" ("or exemplary problem solutions") and Wittgenstein's "examples": see Kuhn (1970a) and (1974) (both written in 1969, a crucial year for Kuhn's philosophical development), and Wittgenstein (1953), part I, §71, §77 and §133.

At the heart of this view lies a dangerous form of imperialism, according to which disciplines and their practitioners must conform: they must not judge one another, and they must not try to describe a common world in collaboration with other disciplines, since each one has its own. Kuhn's philosophy, regardless of its author's intentions, tries to convince us that no fundamental decision or responsibility on our part is required, that we really should not waste our time on criticism or on trying to understand, and that everything will and must go well if we only fall into step behind the institutionalized scientific dogma of the day and its experts. The risk is to replace philosophical and scientific values of truth, rationality, and the freedom of thought with political power, solidarity and commitment to belief. If truth is nothing more than solidarity, then all of our questions are political questions – and the only question that matters is: which side are you on?

Kuhn's relativism gives rise to a sort of conservative defence of whatever belief system is construed as rational according to the established scientific community. Despite revolutionary science is acknowledged, critical attitude is systematically discouraged: instead, normal science is regarded as the essence of the scientific enterprise, and dogmatic commitment to a paradigm (or a lexicon) is upheld as a necessary prerequisite for rational knowledge and social harmony.

Criticism and the quest for truth

By contrast, Popper's philosophy offers an account of how scientific knowledge can be objective and rational without being certain and without grounding itself upon expert opinion, consensus, and the solidarity of belief. By regarding criticism, not description, as the alternative to justification, it offers an account in which truth, and not power, is the key element.

I think that the real difference between Popper and Kuhn is not about the possibility of falsification or the existence of normal science. It is about the role of truth, the value of criticism and the nature of the bond that unites scientists into a community. Popper and Kuhn agree that there is no such thing as an objective criterion for truth, but Kuhn takes this to mean that truth plays no role at all in theory appraisal and theory choice, while Popper maintains that truth plays the role of a regulative idea. As a consequence, Kuhn characterizes the bond uniting scientists in terms of shared beliefs: since it is not possible to prove the truth of such beliefs, scientists can not help but uncritically commit themselves to them. Popper, on the other hand, characterizes this bond in terms of the search for truth, believing that only truth and the critical attitude enable a scientific community to be an open society.[40]

[40] See also Notturno (2000), chapter 10, especially pp. 238-239.

Institutionalism might be a well-intentioned response to the collapse of foundationalism. But it is an inadequate and perhaps even dangerous response. It is well-intentioned because it is meant to oppose irrationalism, but it is inadequate because it fails to rid itself of the demand that we justify our beliefs. And it is dangerous because such a demand, in the absence of our traditional ideal of truth, can easily lead to an even more authoritarian approach to science than the infallibilist one that it was supposed to replace. And authority is authority – regardless of whether it is the authority of the Pope, the authority of experience, or the authority of the scientific institution. And it is useless as proof to anyone unwilling to defer to it.

What is disappointing of logical positivism is the same thing that is disappointing about Kuhn's view of science as an institution: it is the parochial policy of making outsiders of those who criticize the insiders too sharply; it is the obscurantism that rejects alternative theories as meaningless (as opposed to simply as false); and it is the authoritarianism that insists that such theories, if not actually meaningless, can not be rationally discussed. All this fits very well with the policy of never admitting that you are wrong and that your opponents are right.[41] But what is really unacceptable about both Kuhn and logical positivists is that community, politics and power all become far more important than truth.

As soon as we give up the idea that we can show that our theories are true, the epistemic problem becomes the problem not of what to believe, but of how to question. It becomes the problem of how seriously question statements that we can reasonably believe to be true.

Popper's philosophy offers a middle way between two opposite, irrational attitudes, dogmatism and relativism. It offers an account of how scientific knowledge can be objective and rational without being certain, without appealing to induction, and without grounding itself upon expert opinion, consensus, and the solidarity of belief.

Falsifiability is the advice that we should be aware of our fallibility and critical of our theories. But we are able to move from the awareness of our fallibility to criticism of our theories only if we are looking for the truth. That is why truth, on Popper's account, is still the regulative idea of scientific inquiry and rational discussion.

---

[41] On these issues, see Joseph Agassi's remarks in his (1997) and (2002); the latter paper, in particular, deepens and substantiate the considerations I am here only outlining.

Rationality without foundations

Popper thought that Carnap was working from an entirely different view, according to which science has a special authority and scientists have the right to speak with authority because they appeal to facts in order to verify their theories and are thus the ones who really know. Lakatos and Kuhn both rejected this view, but only to replace it with one in which the scientific community could protect its theories – and hence its authority – from criticism without justifying them at all. The basic difference between Popper and his critics, in each case, is whether the objectivity and rationality of scientific knowledge results from exposing our theories to contrary evidence or from insulating them against the possibility of refutation. And the basic difference between their methodologies, when it comes to objectivity and rationality, is whether and to what extent they attempt to protect scientific theories – and, indeed, the authority of science itself – from criticism.

While traditional philosophy regarded the justification of beliefs as the goal, and the critical method a way of reaching it, to Popper the goal is truth, and it is the method of investigation itself that is rational.[42] This is the core of the revolution Popper marked in twentieth century philosophy of science – rationality requires no foundation, only critical dialogue: it is the end of foundationalist philosophy.

The fundamental epistemological fact of twentieth century was the collapse of foundationalism. Its fundamental epistemological problem has been how to respond to it. Some philosophers have concluded that scientific knowledge is unjustified and hence irrational after all. Others – indeed, the majority – have opted for retaining the idea that scientific knowledge is justified, but for weakening either the idea that truth is correspondence with reality or the idea that justification shows that a statement is true. Each of these responses retains the foundationalist theory of rationality, according to which it is irrational to accept a belief that has not been justified and obligatory to accept one that has. Wittgenstein's idea that science is grounded in a form of life; Carnap's idea that it is grounded in the external questions of a linguistic framework; Kuhn's (similar) idea that it is grounded in the acceptance of a scientific paradigm, or lexicon; Rorty's idea that it is grounded in the solidarity of community – each of these is a return to Hume, since the fundamental idea of each of them is that our knowledge is grounded, and must be grounded if we are to regard it as rational. On the contrary, Popper's proposal is that of a rationality without foundations.

---

[42] Truth (*i.e.* correspondence) is an ideal, a regulative idea, always sought and never sure to have obtained: indeed, we may well obtain truth, but not justified certainty that we have obtained it. The linguistic turn no longer threatens realist metaphysics and objective science.

The collapse of foundationalism has no implication whatsoever for truth. It does not show that truth does not exist, it does not belittle it and it does not reduce it to solidarity and consensus. It only shows that – as Socrates beautifully said – we are living in the twilight zone between knowledge and ignorance, where the views that we hold may be true, but where we are unable to know that they are. The failure of foundationalism is not the failure of the reach. The twilight zone between knowledge and ignorance is an ideal well worth holding on to.

References

AGASSI, Joseph. 1997, "Let a Thousand Flowers Bloom: Popper's Popular Critics", Anuar, VII, 1997, pp. 5-24.

———. 2002, "Kuhn's Way", Philosophy of the Social Sciences, 32, 2002, pp. 394-430.

BARTLEY, William W., III. 1990, Unfathomed Knowledge, Unmeasured Wealth: On Universities and the Wealth of Nations, La Salle, Illinois: Open Court, 1990.

FEYERABEND, Paul K. 1970 "Consolations for the Specialist", in Imre Lakatos, Alan E. Musgrave (eds.), Criticism and the Growth of Knowledge, Proceedings of the International Colloquium in the Philosophy of Science, London, 1965, volume 4, Cambridge: Cambridge University Press, 1970, pp. 197-230.

———. 1978 "Kuhns Struktur wissenschaftlicher Revolutionen. Ein Trostbüchlein für Spezialisten?", in Paul K. Feyerabend, Der Wissenschaftstheoretische Realismus und die Autorität der Wissenschaften: Ausgewählte Schriften, Braunschweig: Friedrich Vieweg & Sohn, 1978, pp. 153-204.

FULLER, Steve W. 2000, Thomas Kuhn: A Philosophical History of Our Times, Chicago-London: The University of Chicago Press, 2000.

HOYNINGEN-HUENE, Paul. 1989a, Die Wissenschaftsphilosophie Thomas S. Kuhns: Rekonstruktion und Grundlagenprobleme, Braunschweig: Friedrich Vieweg & Sohn, 1989; English translation by Alexander T. Levine, Reconstructing Scientific Revolutions: Thomas S. Kuhn's Philosophy of Science, Chicago-London: The University of Chicago Press, 1993.

———. 1989b "Idealist Elements in Thomas Kuhn's Philosophy of Science", History of Philosophy Quarterly, 6, 1989, pp. 393-401.

KUHN, Thomas S. 1962 The Structure of Scientific Revolutions, Chicago-London: The University of Chicago Press, 1962; second edition, 1970.

———. 1970a, "Postscript–1969"in Kuhn (1962, 1970), pp. 174-210.

———. 1970b, "Logic of Discovery or Psychology of Research?", in Imre Lakatos, Alan E. Musgrave (eds.), Criticism and the Growth of Knowledge, Proceedings of the International Colloquium in the Philosophy of Science, London 1965, vol. 4, Cambridge: Cambridge University Press, 1970, pp. 1-23.

———. 1974, "Second Thoughts on Paradigms", in Frederick Suppe (ed.), The Structure of Scientific Theories, Urbana-Chicago-London: University of Illinois Press, 1974; second edition, 1977, pp. 459-482.

———. 1989, "Possible Worlds in History of Science", in Sture Allén (ed.), Possible Worlds in Humanities, Arts and Sciences: Proceedings of Nobel Symposium 65, Berlin: Walter de Gruyter & Co., 1989, pp. 9-32.

———. 1991, "The Road Since Structure", in Arthur Fine, Micky Forbes, Linda Wessels (eds.), PSA 1990: Proceedings of the 1990 Biennial Meeting of the Philosophy of Science Association, East Lansing, Michigan: Philosophy of Science Association, 1991, vol. 2, pp. 3-13.

———. 1992, The Trouble with the Historical Philosophy of Science, An Occasional Publication of the Department of the History of Science, Cambridge, Massachusetts: Harvard University, 1992.

———. 1993, "Afterwords", in Paul Horwich (ed.), World Changes: Thomas Kuhn and the Nature of Science, Cambridge, Massachusetts-London: The MIT Press, 1993, pp. 311-341.

NOTTURNO, Mark A. 2000, Science and the Open Society: The Future of Karl Popper's Philosophy, Budapest: Central European University Press, 2000.

SANKEY, Howard 1997, Rationality, Relativism and Incommensurability, Aldershot: Ashgate, 1997.

WITTGENSTEIN, Ludwig 1921, "Logisch-philosophische Abhandlung", Annalen der Naturphilosophie, 14, 1921, pp. 185-262; English translation by David F. Pears and Brian F. McGuinness, Tractatus Logico-Philosophicus, London: Routledge & Kegan Paul, 1961.

———. 1953, Philosophische Untersuchungen, edited by G. Elizabeth M. Anscombe and Rush Rhees, Oxford: Basil Blackwell, 1953.

———. 1969, On Certainty. Über Gewißheit, edited by G. Elizabeth M. Anscombe and Georg H. von Wright, Oxford: Basil Blackwell, 1969.

CHAPTER FOUR: STEPHANOS CHEROUVIS

ONTOLOGICAL COMMITMENTS IN CHOMSKYAN LINGUISTICS

Introduction

This paper deals with some ontological issues in Chomskyan Linguistics. I will argue that the lack of proper ontological debate in this paradigm has given rise to misinterpretations and has made the Chomskyan enterprise vulnerable to attack.

Chomsky and metaphysics

A paper that focuses on Chomsky's ontological commitments might sound a little odd to some people, especially those familiar with Chomsky's linguistic project over the past 40 years. And I say that, because Chomsky has frequently stated that his theoretical constructs carry no "metaphysical import", since we have no clear understanding of notions such as "mind", "matter", "body", "VP" (the part of a sentence S that constitutes a phrase and includes a verb) etc; we use such terms only to refer to certain aspects of the world. He has also mentioned a number of times that he finds metaphysical and ontological questions irrelevant and a form of harassment.

As far as the first statement is concerned (that theoretical constructs such as VPs carry no metaphysical import), some would consider it a metaphysical claim in its own right. Furthermore, when we insist that our theoretical constructs pick up things in the world (Chomsky maintains that this is the case in his work), our investigation has to do with some aspects of reality, and this is what metaphysics is all about!

I will be using the terms "mind" and "mental" here with no metaphysical import. Thus I understand "mental" to be on par with "chemical", "optical" , or "electrical". Certain phenomena, events processes and states are informally called "chemical" etc., but no metaphysical divide is suggested thereby. The terms are used to select certain aspects of the world as a focus of inquiry We do not seek to determine the true criterion of the chemical, or the mark of electrical, or the boundaries of the optical. I will use "mental" in the same way with something like ordinary coverage, but no deeper implications (Chomsky 1995, p.1).

I would also consider the second statement (the fact that ontological questions are a form of harassment) a compliment, at least from the point of view of philosophy. When philosophical questions or reservations (regarding metaphysical implications, for example) harass rather than fall on deaf ears, it means that philosophy still has a prominent role in science. However, Chomsky is right when he feels that he has been harassed by those (I suspect) who demand an explicit ontological account of his theory that would include assigning an ontic status to certain constructs, something that would, of course, make an attack on him easier. But this is bad metaphysics! Certainly not the one that I believe most of us who are genuinely interested in it have.

Metaphysics - which has traditionally been held to be the root of all philosophy - is the systematic investigation of the fundamental structure of reality. It includes, as an important sub-division, ontology: the study of what categories of things do or could exist. (Lowe 2000, p.4).

What is language

Let us now turn to what is that thing called language according to Noam Chomsky.

Chomsky's conception of language involves a distinction between what he calls the I-language and the E-language. The externalised language (E-language) corresponds to the set of sentences or utterances that are external to the mind of the language speaker and has been the focus of inquiry of many schools, including the Behaviourists, prior to the emergence of the Generative/Transformational approach in linguistics that Chomsky has been associated for the past 40 years. On the other hand, the internalised language (I-language) is a view of language that considers it an intrinsic part of the mind of the person who knows it.

Chomsky considers language (that is I-language) to be an element of the natural world. He holds the same for concepts such as "mind" and "mental". These elements are the focus of his unique version of naturalistic inquiry. I call it unique, because by naturalism, Chomsky refers to what he calls methodological naturalism, which is set against a certain doctrine of methodological dualism. The later advocates that language and the mind should not be approached in the way that the natural sciences study natural objects (Chomsky 1995, p.28). As for the case of naturalism in the study of language and mind outside linguistics, the view that one should attempt to account for such phenomena purely in terms of the physical or material properties of the human body, Chomsky has indicated that such an approach seems problematic, since we lack some concrete notions for "body", "physical" or "material" (Chomsky 1995, p.37). He calls this approach metaphysical naturalism. But one does not need to have a "final" account of the "mental" or the "physical" in order to engage in in some metaphysical analysis (of a naturalistic short, for example). Most scientists consider the ontological implications of their research on the basis of the empirical evidence available to them. This is something that linguists within the old Generative tradition refuse to accept. And it has resulted in the absence of a proper debate on the metaphysical or ontological implications of linguistic theories[43]

Ontological status of the I-language

I will return now to the notion of the I-language and its ontological status. Chomsky's I-language is located in time and space and is partly determined by the biological make-up of the brain along evolutionary lines (in other words, it is partly innate). This is his answer to what is known as Plato's problem; that is, how complex knowledge of language arises in the mind, given the minimal nature of stimuli that are available to children during the critical period of language acquisition. Furthermore, we have also empirical access to it through the speaker/hearer's intuitions.

Entities such as VPs and NPs that are part of our I-language are descriptive of an extra-theoretical reality, and this is the psychological reality of the human mind/brain. According to Chomsky, language should be viewed as a psychological mechanism whose entities, such as NPs and VPs, are somehow represented in the mind of the speaker.

So far, a psychological mechanism is the best way to account for the phenomenon, although one should hope for a possible explanation of language along theories that focus on the physical properties of the brain. Indeed, we have evidence for the existence of physical mechanisms underlying the use and acquisition of language. Experimental evidence has shown (Smith 1999, p93) that PET (Positron Emissions Tomography) scans indicate an increase in the flow of blood in specific parts of the brain, when someone performs certain linguistic tasks (such as sentences in passive voice, for example) under experimental conditions. However, we are far from anywhere near to an adequate purely materialistic approach to language. And we may never be, according to Chomsky and others. It is possible that an explanation of the language phenomenon in terms of neural structures, for example, is beyond the scope of human intelligence, since humans are part of the natural world and subject to certain limitations. Such limitations may be

[43] For some discussions on ontological matters see Carr (1990), where an alternative ontological metatheory is offered for the Chomskyan project, Jackendoff (2002) and Antony & Hornstein (2003).

the reason behind an apparent lack of those concepts needed for a complete understanding of language along such terms. This is, of course, an epistemological theory, regarding the possible limits of human knowledge. But a metaphysical idea is present too: that there is a reality independent of our observations.

The Chomskyan Universals

I will turn our attention now to the notion of Universal Grammar and the different ontological interpretations it has given rise.

The initial part of our I-language is what Chomsky and his followers call UG. "It is a set of linguistic principles we are endowed with at birth in virtue of being human", according to Neil Smith (1999, p.42). This is an attempt to account for the striking similarities that are shared by the world's languages. These similarities are taken to be universals of language. These universals form the initial (biologically determined) cognitive state of the I-language. As part of that initial state of the I-language, they are crucially empirical and have causal properties. They are not concepts in any sense. The same holds, naturally, for all the linguistic entities and constructs postulated by Chomsky. As mentioned above, entities such as NPs are somehow represented in the human mind, bur they are not concepts in the strict sense.

There has been some confusion regarding the matter, with many attributing a conceptualist approach to Chomsky, because of his treatment of universals. According to typical conceptualism, universals are concepts existing in the mind. The term is usually set against ontological realism, which, on the contrary, states that universals exist independently of the mind. Botha (1992) and Katz (2000) have branded the Comskyan approach a form of conceptualism, due to its strong internalism. This supposed conceptualism is set in opposition to the realism of the Katzian model, in which universals are understood as abstract objects in the Platonic sense that would exist even if there were no minds to capture them.

Of course, one cannot blame Chomsky, because others have provided a rather confusing ontological interpretation of his enterprise. But one also must not forget that such confusions are present in most university theatres where contemporary linguistics is taught, and that such confusions trouble many of those (including myself) that are committed to the Chomskyan project.

Causation

Language is in the head, and what goes on in there -in a certain mental linguistic faculty- is greatly determined by our genetic make-up. So, in a sense, we are born with language.

At birth our language faculty is in its initial mode with a working set of fundamental principles. These principles are the same for every human being, regardless the linguistic environment (i.e. English, French etc). We call this biological sate Universal Grammar. The maturing of our linguistic module involves the emergence of a set of parameters that account for the differences among the world's languages. Linguists (especially Chomskyans) consider the phenomenal variation in world's languages insignificant. The language faculty is an elaborate system, and small changes in such a system, perceived as parameters of our mental grammar, can yield phonological, morphological, syntactic etc disparities that seem substantial to the unsuspicious mind. But they are only phenomenal! When the language faculty has attained its full state, with the parameters that give us English for example set, we say that a speaker has acquired the English language as his/her mother tongue. Knowledge of language, therefore, is the coming of a certain linguistic state in the mind/brain.

But we speak and hear language; we come up with utterances of all shorts. In a sense, there is something that is publicly performed, and some might call it a form of behaviour. So is there a linguistic behaviour, and if so is that behaviour caused by the inner workings of the language module in the human mind/brain? In other words, can the Chomskyan paradigm accommodate causation?

Causation is a problematic issue in philosophy of mind. Physicalists, including those who are willing to accept that mental events are supervenient upon brain events, argue that there is no causal influence at the mental level. Causation operates strictly at the physical level: only physical states can cause other physical states.

Chomsky, who holds that we do not know what the physical level is, and we only use the words "physical" and "mental" to refer to certain aspects in the world, is cautious with the issue too, despite that linguists associated with his paradigm, are willing to incorporate causality in their theories. Smith (1999) accepts that causation underlines the relation between grammatical competence (the I-language mental state) and performance. Components and internal structures in the language faculty trigger our articulation and perception of language, as well as our manipulation of concepts and meanings.

Entities like Noun, Verb, Complementizer, and the hierarchical tree structures into which they enter, have causal properties in the sense that they underlie any explanation of the linguistic knowledge and intuitions of native speakers. Thus, the fact that I interpret the him in Bill thinks John likes him as referring either to Bill or to some third person, not to John, is caused by the rules and representations in my head. Different representations would cause me to make different judgements (Smith 1999, p.140)

A number of interpretations can feature here. First, that causation operates at the mental level, which has huge ontological implications for the mental reality . Second, that a causal relation may exist between the mental level of competence and the level of performance. None of these interpretations are to be assessed in this paper. Needless to say that such issues require intensive discussion that possibly includes the introduction of a metaphysical framework.

Chomsky, however, is reluctant to warrant such an interpretation of the relation between what goes on in our mental language faculty and what reaches our ears. Chomsky views I-language as a computational procedure that generates linguistic expressions. These expressions become available to our performance systems (not part of our linguistic module) and surface as a language. We have two types of performance systems: an articulatory-perceptual and a conceptual-intentional. When our I-language grinds out linguistic expressions, these include instructions for the performance systems that are gathered at two interface levels:

We are studying a real object, the language faculty of the brain, which has assumed the form of a full I-language and is integrated into performance systems that play a role in articulation, interpretation, expression of beliefs and desires, referring, telling stories, and so on… It is only by virtue of this integration into such performance systems that this brain state qualifies as a language (Chomsky 2000, p. 27)

It seems that Chomsky is not making a strong case for causation. Our linguistic performance is not caused by our mental grammar. Rather our independent performance systems access certain information at interface levels, and that brings about language.

There are no strong grounds for believing that behaviour is caused, at least in the sense of 'cause' that we understand. I know of no good reason to suppose that the Cartesians were wrong in their basic descriptive observations about behaviour, in particular, about the creative aspect of language use; that is, behaviour that appropriate and coherent but uncaused, the normal case, they argued. It could be well that we are approaching mysteries-for-humans, at this point (Chomsky 1994, p.162)

Conclusion

I think that what is evident here is that the lack, not of ontological commitment, but of ontological debate has done great damage to the Chomskyan undertaking. The problem of "forced" ontological indeterminacy, as many critics of Chomsky call it (Carr 1990 and again Botha 1992) has hinder progress in the field. The continuous absence of such a debate ignores the fact that "the pursuit of metaphysics is inescapable for any rational being" (Lowe 2000, p.4).

Bibliography

Antony, L., M. & Hornstein, N. (ed) (2003) Chomsky and his Critics. Blackwell.

Botha R., P. (1992) Twentieth Century Conceptions of Language. Oxford UK & Cambridge USA: Blackwell.

Carr, P. (1990) Linguistic Realities. New York: CUP.

Chomsky, N., 'Chomsky Noam', in S.Gutteplan (ed.) (1994) A Companion to the Philosophy of Mind. Oxford: Blackwell, pp153-67.

———. (1995) Language and Nature. Mind 104:1-61.

———. (2000) New Horizons in the Study of Language and Mind. CUP.

Jackendoff, R. (2002) Foundations of Language. Oxford: Oxford University Press.

Katz, J., J. (2000) Realistic Rationalism. Cambridge: MIT Press.

Lowe, E., J. (2000) An Introduction to the Philosophy of Mind. CUP.

Smith, N. (1999) Chomsky: Ideas and Ideals. CUP.

CHAPTER FIVE: BIAGIO TASSONE

HUSSERL'S LOGICAL INVESTIGATIONS: THEORY OF SCIENCE AND THE 'BREAKTHROUGH' TO PHENOMENOLOGY

Within his Logical Investigations[44] Husserl first formulated in detail his conception of pure logic as a 'science of science' and of phenomenology as the foundation for this theory of science. According to Robert Sokolowski, "the Logical Investigations begin and end with the problem of evidence and fulfillment in the context of scientific theory" (Sokolowski, 1977, p.105). Arguably however, it is precisely Husserl's rather obscure and quite ambitious theory of science that many contemporary readers of the Logical Investigations may be most unfamiliar with. In the present paper we shall attempt to show that, within the Investigations, Husserl was, indeed, not concerned with logic in its formal sense alone nor with thought as constituting logical forms exclusively but rather with the interaction of the subjective 'sense bestowing' cognitive acts and their unique relation to what he called pure logic [reine Logik]. This concept of a pure logic- formulated non-psychologistically- is initially introduced as the foundation of a very broadly conceived philosophy of science. For Husserl, clarity on logical matters was an essential pre-requisite of a fruitful engagement with epistemology and the attempt to gain insight into a genuine understanding of science.

In what follows, we shall proceed by examining Husserl's Logical Investigations with a special emphasis on the Prolegomena to Pure Logic in an attempt to extract his early theory of science. The main point to be made is that this early philosophy of science rests on an ambivalent foundation. On the one hand, knowledge of the pure realm of ideal objects is said to constitute the basis upon which the sense of all theory is supposed to be found. Simultaneously, an explicit reference to a more formal notion of

---

[44] E. Husserl, *Logische Untersuchungen*, Tübingen: Max Niemeyer, (Verlag, 1913, 1921). Critical edition: *Husserliana*, Vols. 19 (ed. by Ursula Panzer) & 20 (ed. by Ullrich Melle). Translated as *Logical Investigations*, in 2 volumes by J.N. Findlay, latest edition with a new preface by Michael Dummet and Introduction by Dermot Moran. (Routledge, 2000). All English quotations will be from the latest edition, henceforth referred to as (LI) followed by page number and relevant German section numbering.

logical consistency, as found in 'existence proofs' for mathematics, is also alluded to. Within the present paper the possible reasons for this tension in Husserl's early philosophy of science will be subsequently examined and some of its consequences explored.

Logical Investigations

In later life, Husserl claimed that the Logical Investigations were the product of a "ten year long effort to clarify the Idea of pure Logic by going back to the sense bestowing or cognitive achievements being affected in the complex of lived experience of logical thinking" (Mohanty, 1977, p.197).[45] In his Logical Investigations (1900-1901) Husserl gives an outline of what he calls "pure-logic": the theoretical foundation of all logical and scientific knowledge. This is paralleled by a three-level theory of formal logic and an attempt to ground both in a phenomenological theory of cognition. This elaboration is the culmination of Husserl's early studies of mathematics and logic. For the purposes of a study of Husserl's philosophy of science however, the theory of pure logic and the theory of manifolds (which ground it) are the most important aspects of the Investigations. The text itself is, as Dermot Moran observes, "extraordinarily rich in philosophical insights and rigorous conceptual analysis" (Moran, 2000, p.164).[46] But while it is true that many valuable philosophical insights are to be found between its covers it is also true that the ideas are not always neatly organized or clearly presented.[47] This fact alone makes the Logical Investigations a difficult work to clearly summarise.[48] For our purposes, within the context of the present paper, we shall deal with only some of the diverse ideas presented within the Investigations and we shall attempt to take account of them thematically. Our

---

[45] The quote comes from Husserl's essay "The Task And The Significance of the Logical Investigations" from Husserl's *Phänomenologische Psychologie* lectures of 1928, reprinted in Husserliana Volume IX and edited and translated in J.N. Mohanty's (editor), *Readings on Edmund Husserl's Logical Investigations*, Martinus Nijhoff, (The Hague, 1977).

[46] See his article "Husserl's Critique of Brentano in the *Logical Investigations*" Manuscrito XXIII, pp.163-205, October. 2000.

[47] Something Moran also notes in his article, i.e., "...the immense length, density...complexity...convoluted structure and meandering thought paths" (ibid.)

[48] Although this fact has not prevented previous scholars from trying, see for example Marvin Farber's *The Foundation of Phenomenology: Edmund Husserl and the Quest for a Rigorous Science of Philosophy*, SUNY Press, Albany (1943, rev.1967). Farber paraphrases the (at the time, untranslated) *Investigations* using 200 or so pages of his study to do so.

main interests will be the themes that centre on Husserl's idea of science and the most important of these are, respectively: the status of ideal objects, the phenomenological account of the grasping of universals, the formulation of the theory of 'pure logic' and the theory of manifolds underlying it.[49] Lastly, we will, in the course of this paper, comment on some of the epistemological insights and ontological views presented in the Logical Investigations by critically exploring, what will be called, the "species" theory of meaning maintained by Husserl in his text. Using examples from the philosophy of mathematics we shall attempt to show how the species theory of meaning is unsound and why Husserl was prompted to abandon it in his later writings. Finally, we hope to show how all of the above views after being elaborated expanded Husserl's own understanding of what constitutes true or genuine science. I hope to try and make clear how Husserl came to view genuine science as grounded in lived experience the foundations of which are to be explored through a phenomenological analysis of consciousness. Husserl called the Investigations his 'breakthrough' text into phenomenology[50] and in many ways this was true. The basic aporias outlined in the Investigations and many of the procedural methods put fourth to help overcome them were to remain with Husserl for the rest of his philosophical life.[51]

---

[49] Husserl's conception of notions of perception, intuition, judgement, and also the role of 'meaning intending' and 'meaning fulfilling' acts will be looked at in passing however not dealt with in any detail.

[50] See the forward to the second edition where Husserl writes, "My *Logical Investigations* were my 'break through' not an end but rather a beginning" (LI, Vol. I, introduction, p.3).

[51] While it is true that Husserl's thought changed considerably after the publication of the *Ideen I* in 1913, i.e. twelve years after publication of the *Investigations*, nonetheless, among the conceptual tools that he retained, and that remained central to his mature phenomenology, way may list the following: the notion of immediate 'Evidence' and intentional fulfilment of mental acts as the basis of his epistemology, the phenomenological notion of truth developed in the sixth Investigation, and finally, some scholars maintain that there is an early, albeit unthematised, version of the later 'transcendental' or 'phenomenological' reduction at play in the *Logical Investigations*. Cf. Quentin Smith "Husserl's Theory of the Phenomenological Reduction in the Logical Investigations" *Philosophy and Phenomenological Research*, Vol. 39, No. 3. (Mar., 1979), pp. 433-437. Smith claims that this early reduction remains confined to the empirical ego, reducing it to its essential functions for better descriptive analysis. This strategy may have been inspired by the insistence within Brentano's descriptive psychology to reduce the field of study to only psychic phenomena thereby bracketing out genetic concerns and all speculation as to the ontological status of the psychical and physical phenomena (cf. Brentano, 1995, pp.)

Pure Logic as a 'science of science'

The early chapters of the Prolegomena to Pure Logic, which begin the Logical Investigations, introduce an innovative formulation regarding the nature of logic. Distinguishing between logic as a normative, practical and theoretical science, Husserl claims that pure logic is in essence a theoretical discipline however one that may be reformulated to correspond to the organisation of both normative and practical sciences. The true nature of logic, according to Husserl, is to be a theory of all possible theories and also to help explain how different scientific theories interact. Thus, pure logic cannot be identified with any individual normative discipline or with any individual normative technology (such as, e.g., the scientific method as it is followed in the empirical or natural sciences) instead 'pure logic' is said to be intimately related to meaning. Husserl writes:

Pure logic, wherever it deals with concepts, judgements, and syllogisms, is exclusively concerned with the ideal unities that we here call 'meanings'. If we take the trouble to detach the ideal essence of meanings from their psychological and grammatical connections, if we try, further, to clear up their a priori relations of adequacy, founded in this essence, to the objective correlates that they mean, we are already within the domain of pure logic (LI, Inv. I, p.224, § 29).

Husserl claims that logical laws have in themselves no essential relation to empirical processes of thinking (this was the error of 'psychologism' prevalent at the time Husserl was writing)[52] but logical laws do allow for normative reformulations. Logic, for Husserl, we can say, is a kind of special theoretical science. Logical laws apply to all the other existing sciences; chemistry, physics, biology, etc. Logic governs the procedural methods of these sciences in their goals of attaining knowledge, however, logic also and simultaneously governs itself.[53] The last chapter of the Prolegomena volume of the Investigations introduces this most important theme in Husserl's philosophy of science at this stage, namely, the concept of pure logic as a science of science or as mathesis of all possible theory. Logic, after Husserl's

---

[52] For an excellent overview of the problematics involved in the 'Psychologism' debate within German philosophy (however from a Sociological perspective) cf. Martin Kusch's, *Psychologism: A Case Study in the Sociology of Knowledge* (Routledge, 1995).

[53] This notion of logic as guiding the sciences was also shared by Frege; see for example his essay "Logic" in Gottlob Frege *Posthumous Writings* edited by, Hermes, Kambartel & Kaulbach, translated by Long & White. (University of Chicago Press, 1979), pp.126-152 and "Thoughts" in *Collected Papers on Mathematics, Logic, and Philosophy*. Edited by Brian McGuinness (Basil Blackwell, Oxford, 1984), pp.351-372.

initial investigations into it, was always viewed as a theory of science,[54] however, it was only within the Investigations that a sustained presentation of this new conception of logic is undertaken. Logic as the science of form is viewed by Husserl as supremely general in its referential power and applicability. We have claimed above that, for Husserl, formal logic deals with both meaning and science. Logic understood as a theory of science will therefore mean, on this view, that there is no science that is not covered by the laws of pure logic. What this claim amounts to is, basically, that as logic is the most universal science and because it is applicable to whatever is a science [one could say to science in general] including itself, there neither is nor can there exist a particular science that is not governed by the laws of logic. In other words, pure logic is the theory of science par excellence- it is a 'science of science'.

We will explore the notion of what precisely a 'theory of science' should consist of in a moment. First we will note that the notion of a 'pure logic,'[55] before the concept is refined in later writings, stands as a three-tier conception.[56] Husserl writes of the "three sets of tasks" that must be distinguished before a thorough understanding of the role of pure logic may be elucidated (LI, Prol: p.152, §67). In a sense, it would seem, what is called 'pure logic' can itself be broken down into various sub-branches or levels. For example, the first level consists of apophantics (as Husserl will call it in his later text: Formal and Transcendental Logic), these are the truths studied by the logic of statements or propositions (as the name implies). These propositions or statements are composed of essential meanings and their combinations and are therefore said to determine what it means to make-sense, i.e. to convey meaning (such as for example in the working out of regulatory or general grammatical rules so as to avoid contradiction or nonsensical expressions in natural linguistic expression). Husserl states that these first or primitive concepts 'make possible' "the interconnected web of

[54] See his early writings on logic and mathematics collected in E. Husserl, Collected Works Vol. V: *Early Writings in the Philosophy of Logic and Mathematics*, translated by Dallas Willard, (Kluwer Academic Publishers: Dordrecht, 1994), The Willard translation is made up of documents mostly taken from Husserliana Volume XXII, *Aufsätze und Rezensionen* (1890-1910) edited by Bernhard Rang, (Matinus Nijhof, The Hague), 1979. See also Husserl's first book *Philosophie der Arithmetik. Mit ergänzenden Texten (1890-1901)*. Edited by Lothar Eley. (Martinus Nijhoff), 1970. Translated as: *The Philosophy of Arithmetic* in E. Husserl, Collected Works Vol. X, translated by Willard, (Kluwer: Dordrecht, 2003).

[55] And this term should here be understood as signifiying a purely theoretical logic, i.e. logic distinguished from normative and practical applications.

[56] See Bernet (1993), pp. 44-51 for an excellent account of the idea and task of pure logic.

knowledge" (ibid, p.153). On this apophantic level- what is also called, the level of the pure categories of meaning- logic deals with the forms of different meanings and how these fit together within the frameworks of different theories. The second level of pure-logic has to do with the 'things' which the above 'apophantic statements' can refer to. This second level analyses things (objects) and also what are called "states-of-affairs" (Sachverhalte) objectively. It covers all existing phenomena and their relations in their various complexes and other configurations.[57] The different levels of pure logic, functioning as formal theories of meaning, correspond to equivalent levels of what Husserl calls formal ontology.[58] The category of 'thing' or 'object', 'something in general', and the interaction of these objects in a purely formal sense (under an operative category like 'causation' for example) would fall under the sphere of what is called formal ontology by Husserl.[59] Thus, the sciences, on this view, have their corresponding regions of objects to be explored. This is true of both formal and practical sciences. In Husserl's philosophy of science, however, this will not include his later foundational and eidetic science of phenomenology-referred to as 'noetics' in the Prolegomena.

Individual sciences, such as for example factual sciences like geography or biology, are not formal ontologies even though they purport to deal with objects. For Husserl, the approach each science takes toward its object region, or domain, differs and in effect determines the nature of the given

---

[57] In the later *Ideas* texts, Husserl will introduce the separate study of regional fields in transcendental phenomenological analysis. In the *Ideas II* for example, subtitled "Studies in the Phenomenology of Constitution", Husserl evokes a fundamental distinction between the physical (material nature, §§ 1-18), psychophysical (animal nature §§ 19-47) and the cultural/ spiritual realms (spiritual world §§ 48-64). For the later Husserl, it must be stressed, these regions provide the research ground for all positive sciences and the humanistic intellectual disciplines. Later, in *Formal and Transcendental Logic* (1929), Husserl will add the level of 'transcendental logic' and include within its purview the study of the ideal structure of verbal expressions.

[58] This position on logic and ontology will be maintained throughout Husserl's career with only slight modifications, mostly methodological. These are witnessed in the shift from the descriptive psychology of the *Investigations* to the transcendental reflection (the method of reduction) on the structures of mental acts themselves in the *Ideas I* of 1913. The same basic logical structure however is present as the backdrop to the *Ideas* texts, *Formal and Transcendental Logic* and the posthumous work *Experience and Judgement* originally published in 1948.

[59] See, once again, the *Formal and Transcendental Logic* where a much clearer overall presentation of the levels and tasks of pure logic is given. Husserl's foundational part-whole logic would also fall under the category of formal ontology. This part/whole logic will be at the centre of the later *Investigations* and is explored in depth as the main subject matter of Investigation III.

science. In the case of geography, for example, the objects approached: land regions, mountains, rocks and other natural objects, etc. are approached in a general but not in a purely formal way. As formal ontology has no content, i.e. as it is said to give only the pure logical structure of any given region of being. Husserl will claim that different 'material' ontologies are actually what provide the general contents for the sciences and that these material ontologies are themselves subsequently subject to logical structures. In the Logical Investigations we are told that the logical structure of the sciences is such that, given any individual science, the realm or being of that region of objects which that a particular science explores may be dealt with so as to have its subject matter determined in an a priori manner. The general matter, which corresponds to the form of a science as determined by its material ontology, would also establish a clear demarcation of the domain to be studied by any particular given science. For the empiricist, we may note, objects and domains of study are deduced from experiments performed in the laboratory.[60]

In the Investigations, there is a close identification of pure logic with the existing mathematical theory of manifolds[61] in a way that it would serve as a theory of all possible theories. At this stage, this is the third and most general understanding of pure logic. This level of logic was apparently conceived of in an a priori and axiomatic fashion; i.e. as a deductive system that could somehow inform all conceivable or possible theories and that could possibly be realized a priori through a determination of the proper axioms. This last and most controversial element of pure logic was later abandoned. In the Prolegomena, however, Husserl describes pure logic as manifold theory thus:

> [t]he most general Idea of a Theory of Manifolds…a science which definitely works out the form of the essential types of possible theories or fields of theory, and investigates their legal relations with one another. All actual theories are then specializations or singularizations of corresponding forms of theory, just as all theoretically worked-over fields of knowledge are individual manifolds, then all

---

[60] However, later on, in his *Formal and Transcendental Logic*, Husserl will claim that formal ontology and formal apophantics must be very intimately and perhaps even inseparably related (FTL, § 83). Both will be claimed to belong to one comprehensive '*mathesis universalis*' (FTL, §53). Donn Welton points out the similarities between Husserl's approach to science and the early Wittgenstein of the *Tractatus*. Both seemed to share what Welton describes as "the same 'mirroring' of logic and ontology" cf. (Welton, 2000), pp.83-4.

[61] Which should not be understood as any kind of set theory in the contemporary sense; rather, manifold theory in mathematics in Husserl's day referred more to uniquely determined 'regions' that could subsequently be viewed as falling under a theory of a given form. See (Miller, 1982), pp.16-17.

> deductive theoretical work in constructing all actual theories of the same has been done (LI, p.157, § 70).

Because Husserl is interested in simple or very broad logical truths in the Investigations, issues such as, e.g. the complete formalisability of logic or mathematics a la Hilbert is left open. Husserl does not go into questions of Hilbert's program. However, Husserl does seem to imply at times that it is (or may be) viable. This is not surprising, given the period at which Husserl was writing. Today, based on discoveries in logic and meta-mathematics very few logicians would accept the viability of Hilbert's project.[62] However, because formal axiomatization in a strict sense is not at the heart of Husserl's approach to scientific knowledge (and certainly not in the later Investigations and subsequent writings where purely logical studies are forsaken for phenomenological accounts of knowledge and meaning acts) the untenable nature of Hilbert's project therefore certainly does not invalidate all of what the Investigations has to say. It does however point to tensions in Husserl's doctrines at this stage in his thought, some of which we shall take into account in section three below.

We have stated above that, at its primary level, pure logic deals with meaning. Apophantics, we have claimed, deals with statements or propositions and their forms of connection. Apophantics thus sets the standard for which any theory whatsoever must fulfil to be valid as a theory, namely, logical consistency and the non-contradiction of its various composing propositions. Apophantics also roughly corresponds to what we would call formal logic. Ontology, on the other hand, for Husserl deals with logic as expressing the way things are 'in themselves'. On some level (it is implied in Husserl's Investigations) the structure of reality is mirrored by the structure of logic. There is certainly, in the Logical Investigations, no explicitly Kantian attempt to deduce a priori forms of thought into a scheme of categories of the understanding. Rather, for Husserl, (and in a more

---

[62] Gödel's logical work in the 1930's in particular (in essence establishing through his incompleteness theorems, and in an apparently *a priori* way, that there are truths that seemingly transcend all consistent formal systems) certainly make this approach to grounding scientific knowledge much less plausible. For an excellent account, see Ernest Nagel and James Newman *Gödel's Proof* (New York University Press, 1958), and Martin Goldstern and Haim Judah *The Incompleteness Phenomenon: A New Course in Mathematical Logic* (AK Peters Natick, Massachusetts, 1995). The extent to which Gödel's 'Incompleteness theorems' affect Husserl's project is debateable due to the already stated dual nature of the conception of grounded knowledge in Husserlian Phenomenology. Certainly, in his later writings, a purely formal understanding of science is abandoned as phenomenology takes centre stage, yet even in the *Investigations* Husserl conceives of this basically Hilbertian ambition as something to be (hopefully) proven by future research however not a given fact.

Aristotelian vein) in the Logical Investigations our logical concepts are objective and describe not mere phenomena but the ideal validity of the pure objects they are related to. This position may help to account for what has been called the ontological realism (or rather the popular realist readings) of the Logical Investigations. These ontological views however, even at this stage are complicated by the fact that, for Husserl, ontology (or formal ontology) is presented in the Investigations as related to ideal essential objects that are also described as being distinct from metaphysics. For example, metaphysics, Husserl tells us, would ask questions such as, "What, if anything, actually exists?" Furthermore, Husserl claims that:

[Metaphysics] concerns…only such sciences as have to do with actual reality, which does not include all sciences, certainly not the purely mathematical sciences whose objects are numbers, manifolds etc., things thought of as mere bearers of ideal properties independent of real being or non-being (LI, Prol: p.16, § 5).

In chapter one of the Prolegomena, it is claimed that the individual existing sciences are 'theoretically completed' by metaphysics and epistemology (LI, Prol: p.16, § 4). The ideal objects of pure logic on this view are said to determine, "whatever makes sciences into sciences" (ibid). They in turn will be dealt with by this science of science and, it would seem, would 'subsist' even if empirical reality were to cease to exist.[63] Whether Husserl's characterization of what exactly qualifies as "metaphysical" is adequate or not is a subject that will be taken up in due course. First, we will examine more closely Husserl's conception of science in the Logical Investigations.

The grounding of science in the Investigations

It is probably safe to say that pure logic, under the criteria we have outlined above, can be said to fall within the region of metaphysics (although Husserl himself, as we have already noted, never says this). This is especially true insofar as we are told that the true nature of pure logic is intrinsically tied into the existence of 'ideal meanings or objects'. Pure logic, in turn, as we have seen, finds its expression in both normative and practical sciences

[63] Later, the pure 'objects' of logic and apophantics become eidetic singularities and after his transcendental turn in the *Ideas I*, Husserl will carry out a fundamental 'Cartesian' style meditation to prove that absolute consciousness (which now is spoken of as constituting all meaning) would remain as a residuum even after the complete nullifying of the physical world (see *Ideas I*, §§ 33-49). Unfortunately, due to the limitations at hand, we cannot deal in the present paper with Husserl's later thought in any kind of detail.

but the existence of ideal objects is presupposed in the demarcation of the realm of pure logic and so of all sciences as well. This presupposition of the existence of so-called 'ideal entities' in Husserl's phenomenology goes hand in hand with a notion of phenomenology as a science of science which will later be called an eidetic and an a priori science. The purpose of the Prolegomena is to establish the role of pure logic and elucidate its relation to the existing sciences and, as we have claimed, Husserl is unambiguous about the role of pure logic is supposed to play as regards its relation to these sciences.

A difference is pointed out however, at the beginning of the Prolegomena, between science as a practice and science as a logical theory grounding itself and taking account of itself in the sense of providing its own foundations for its existence (LI, Prol: p.17, § 6). Pure logic enters into this foundationalist account of scientific knowledge precisely by establishing the foundations of any kind of knowledge whatsoever in the basic validity or sense of its ideal 'objects', that is, in the sense of its meanings and of the general laws that account for its unity. Husserl writes:

The interests of pure theory require us to treat what constitutes a unified whole of theory in a theoretically closed manner, and not as a mere aid towards external ends (LI, Prol: p.141, §61).

The "external ends", the things or existing objects under investigation by the theory or particular science (and pragmatic concerns such as perhaps the organization of research) are said to be given together with, and mutually inseparable from, an "ideal interconnection of truths" (LI, Prol: p.144, § 62):

What holds of single truths, or single states of affairs, plainly also holds of interconnections of truths or of states of affairs…the interconnections of truths differs from the interconnections of things, which are truly in the former; this at once appears in the fact that truths which hold of truths do not coincide with truths that hold of things posited in such truths (ibid, pp.144-145).

This is most likely the reason why, in the Prolegomena, Husserl claims that: "Epistemology must, of course, not be understood as a discipline that follows upon- or coincides with- metaphysics, but rather one which precedes metaphysics as it precedes psychology and all other disciplines" (ibid). On our reading, this claim is contentious (especially as it is formulated by Husserl here in these pre-transcendental phenomenological writings) however before showing why and attempting any sustained criticism let us first more clearly outline the idea of how science is said to be grounded within the Investigations.

Science, according to Husserl, and in its etymological sense, is concerned with knowing, however; "this does not mean that it itself consists of a tissue of acts of knowing" (LI, p.16, § 6). To understand Husserl here we must take into account his conception of knowledge as an intended versus fulfilled meaning act (a distinction of major importance to the whole of the Investigations as well as his later phenomenology) as well as the special conception of evidence that pertains to true knowledge.[64] The notion of evidence is one that runs throughout the Investigations and it is maintained as being essential to understanding all aspects of truly grounded knowledge. The theory of evidence is correlative to the above notion of intentional fulfilment and is also referred to as the "final authority in all questions of knowledge" (LI, Inv.I, p.230, § 31). Evidence is said by Husserl to be "the immediate intimation of truth itself" (LI, Prol: p.17, § 6). For Husserl knowledge and scientific knowledge in particular may be viewed as a process of classification; that is, of particulars falling under apophantic concepts. The fulfilment of the intentional object within an act of knowledge is Husserl's basic understanding of what constitutes evidence for classification. For Husserl all knowledge involves what we can call a 'synthesis of fulfillment'. In the Logical Investigations, it is explained how the synthesis of fulfilment involves a bringing together of the meaning (i.e. of the intention on the level of the knower) and of the corresponding 'givenness' of the object of knowledge, which in any genuine knowledge is given precisely in its intuitive presence.

In the Prolegomena, Husserl writes: "we separate [true knowledge] off from baseless opinion, by pointing to some 'mark' of the presumed state of affairs or for the correctness of the judgement passed by us" (LI, Prol: p.17, §6). The mark corresponds to the available evidence. This concept of knowing or knowledge is altogether said to be possessed of a dual nature:

> Knowledge in the narrowest sense of the word is the being inwardly evident that a certain state of affairs is or is not, e.g. that S is P or that it is not P. If it is evident that a certain state of affairs is probable to this or that degree, then we have knowledge in the strictest sense of such a probability, but in regard to the being of the state of affairs itself, and not of its probability, we only have knowledge in a wider modified sense (ibid. p.18).

This dual nature of knowledge seems to correspond to the two principles of unity pertaining to scientific theory, namely what Husserl calls the 'essential' and the 'extra-essential' unities respectively, but a similarity exists also to the 'ideal' versus 'real' distinction (inherited from Lotze)[65] which pervades the Investigations. Husserl writes of degrees of probability and of

---

[64] That is, Husserl's contention that all true knowledge (as distinguished from blind belief) can be traced back to the justification or validation given to successful insight by evidence (see the *Prolegomena* pp.18-23) for details.

[65] Cf. note 29 below.

greater and lesser degrees of knowledge. Knowledge, in what Husserl terms, the "pregnant sense" (absolute self-evidence): "counts as the absolutely fixed, ideal limit which the graded probabilities for the being-P of S approach asymptotically" (ibid). At this stage in Husserl's thinking this 'absolute self-evidence' or pure (what Husserl also calls 'indubitable' or 'apodictic') givenness of knowledge can only be achieved within the realm of the psychic acts.[66] So the validations of the psychic acts that convey the 'pregnant' sense of knowledge come first; they are primary (although only investigated in detail much later on in the actual Investigations). However, this is only part of the methodology or procedural mode of science. Equally essential, for Husserl, are what are termed auxiliary devices. Borrowing on his earlier work these are mainly accounted for by signs. Language as a sign system will be said to be indispensable to obtaining and communicating knowledge, however Husserl suspends detailed analysis of language until he begins the actual Investigations proper.[67]

In the Prolegomena, we are told that the individual sciences, as theoretical wholes, are called "systematic unities [systematischen Einheiten] of this or that sort" (LI, Prol: p.24, §10) however they are [all] held together by theory of science itself in the guise of pure logic. Where Husserl writes "systemic" one can read "correlated". Statements or propositions fit together to form theories, the scientific status of these theories in turn seems to be connected to the logical ordering or correlation of the propositions. Here we see reiterated again the ontological issues which featured prominently in Husserl's overcoming of psychologism. A theory of science for Husserl is about, what are called in the Logical Investigations, 'grounded validations' (Begründungen) or what perhaps could better be called 'proper foundations'. A theory of science should be a theory of grounded validations obtained both by means of proper knowing (intuitively grasped direct insights) and by

---

[66] In the first edition of *the Investigations* this is achieved by a special 'psychic' act of reflection within the realm of descriptive psychology. Later (both in subsequent editions of the *Investigations* and in later writings) Husserl will want to say that the act of reflection is not at all empirical and that it can be achieved only by a special or *phenomenological* type of reflection. This is what distinguishes the phenomenology of the *Investigations* from the later 'transcendental phenomenology' of the *Ideas* that characterises Husserl's mature thought. Cf. also note 19 above.

[67] See especially Investigation I and IV. Even there Husserl's investigation is controversial. There is, in effect, no linguistic turn whatsoever in Husserl's writings-early or late. Whether this is a point of contention however is debatable and pertains to the emphasis one places on the relevance of the 'linguistic turn' in the first place. For some varying analyses of Husserl's philosophy of language within the *Investigations*, cf. Paolabanchetti-Robino (1997), Bernet (1988), and Sokolowski in Zahavi & Stjernfelt (2002).

means of something that stands surrogate for proper knowing, namely, knowledge via auxiliary devices.[68] These auxiliary devices- amongst which language plays a central role- can accordingly take over the role of grounded validations and still lead to truth (LI, Prol: p.24, §9). This latter form of knowledge, (i.e. knowledge via inauthentic or surrogate symbols or signs) because it is not truly grounded (i.e. not given in its intuitive presence) is called 'blind' by Husserl. It seems to be equivalent to mechanically derived knowledge of the kind obtained, for example, via a calculus of thought using algorithms. Husserl gives the example of calculating machines in the Prolegomena, but we could just as easily substitute a modern day digital computer. Because knowledge, for Husserl, is more than just obtaining results blindly via any simple form of mechanical calculation but also understanding the results achieved, it is implied that the manipulation of symbols can never totally account for knowledge (certainly not for genuine knowledge). Husserl will in the course of later writings attempt to develop a language that is an appropriate expression of authentic thinking however its analysis is beyond the scope of the current paper.[69]

According to Husserl in the Logical Investigations there are three basic kinds or classes of science that must be distinguished from each other. The three basic kinds of science run as follows: The first is called nomological. As the name implies this is a form of 'law obeying' science. On the nomological model of understanding, certain general and basic scientific laws determine, apparently in an exclusive manner, the unity of truth which in turn connects the propositions involved. Pure logic, supposedly, would be an example of a nomological science on Husserl's treatment of it. The second class of science is the concrete. These concrete sciences are given unity not by means of general laws but rather by means of the things, the objects of which they treat in their theories and how these objects fit together. The concrete sciences are many, here we may list, physics, geography, biology, etc. However, a humanistic discipline such as history would also count as a concrete science. Finally, the normative sciences, the third class of sciences, provide unity to the subject matter being studied by means of determining a basic or fundamental norm ['Grundnorm'] (LI, Prol: § 64). Ethics would be the prime example of a normative science its Grundnorm being "the Good"; however aesthetics (the basic or fundamental normative category of which would be "the Beautiful") would also fall under this category.

[68] Husserl's early theories on semiotics or signs dealt precisely with signs as auxiliary devices leading to legitimate knowledge.

[69] Interested readers are directed to the *Ideas* texts and later writings such as *Formal and Transcendental Logic* (1929) and *Experience and Judgement* (1938).

To further the distinction made above between different kinds of necessity within the sciences, at the end of the Prolegomena, Husserl writes about experimental and non-experimental sciences. Theoretical sciences are said to be nomological while experimental sciences are grounded on, or rather presuppose, the theoretical and nomological ones. As a theory is equivalent to a unity of its true propositions and as, in turn, these are founded on the validity of much more general laws, so it is precisely the general laws that are presupposed in the practical and normative activity of research within the framework of whatever sort of hypothetical deductions or empirical experimentation a science undertakes. Of central concern for Husserl always are the ideal conditions of the possibility of a science, i.e. the ideal conditions of the possibility of any theory whatsoever and not the particular application of laws. Husserl is most interested in the formal conditions of what constitutes a theory in the first place and therefore in his writings pays much more attention to ideal conditions rather than empirical ones. This, it should be pointed out, is in stark contrast to the approach of the majority of philosophers concerned with science and scientific knowledge these days.[70] According to Husserl, theories have objective and subjective conditions of possibility and both the objective and the subjective conditions of the possibility of a theory must have 'ideal' conditions that in turn must be met for their actualization to be possible.

For any theory whatsoever some conditions will be specific and others not. In the case of the subjective conditions to be met, how certain concrete psychological acts of thinking fit together (for example) are not specific for fulfilling the ideal conditions of the possibility for science. That is to say, in the Logical Investigations, psychology is distinguished from noetics. Noetics, according to Husserl and in anticipation of his later transcendental phenomenology, has to do with the unity of acts of thinking and not with accounting for them as concrete acts- as psychology would. For Husserl, psychology accounts for the acts of thinking according to the laws of association, in terms of the contiguity and similarity of different psychic acts, physiological mechanisms and qualitative senstations etc. Noetics, on the other hand, is about evidence. Noetics would thus deal with laws that would hold for all or any 'consciousness in general' and not in any particular empirical case.[71] Husserl is quite clear in this distinction of the objective

[70] i.e. the current obsession *a la* Quine, to naturalise epistemology is thus the exact opposite of Husserl's (later explicitly) anti-natural or philosophical-theoretical approach to many of the same issues.

[71] Later on, as noetics becomes explicitly transformed into phenomenology, the distinction between the real and the ideal of the *Investigations* will make way for more subtle distinctions such as the truly *immanent* versus the *transcendent* contents of the mental acts and the much more radical method of transcendental reduction

conditions of the possibility of theory from subjective ones. The objective conditions require noetic and ideal possibilities for the attainment of truth however the reverse is not the case. Theory depends intrinsically on certain objective conditions holding however no empirical evidence can ever contradict, for example, the objective validity of the laws of formal logic. The validity of the basic laws of logic, for Husserl, would hold regardless of whether anyone could actually gain any insight into them or not.

This brings us to the most controversial aspect of Husserl's theory of knowledge, which we have touched upon already and which subsequently is central to this theory of science at this stage. The aspect meant is that between genuine knowledge given in acts and truths-in-themselves made valid on the strength of evidence gleaned through knowledge of the corresponding ideal unity of propositions. The problematic aspect of Husserl's doctrine in the Investigations stems from, as we have already noted, the positing of ideal entities in dealing with so-called "truths-in-themselves". Husserl will want to say that there is a relation between acts of knowledge and 'truths in themselves' and also that that relation is the relation between a universal (something general) and its temporal or particular instantiation. The particular individual act of knowledge gives the knower, by means of what Husserl calls an 'ideational insight', a truth presented as it is instantiated in a valid act of knowledge. This truth is the ideal correlate (also called the meaning correlate) and the objective condition of the subjective act of knowledge. Accordingly, we are told that:

Meanings constitute…a class of concepts in the sense of 'universal objects'. They are not for that reason objects which, though existing nowhere in the world, have being in a 'topos ouranios' or in a divine mind, for such metaphysical hypostatization would be absurd. If one has accustomed oneself to understand by 'being' only real being, and by 'objects' only real objects, then talk of universal objects and of their being, may well seem basically wrong; no offence will, however, be given to one who has first used such talk merely to assert the validity of certain judgements, such in fact as concern numbers, propositions, geometrical forms etc., and who now asks whether he is not evidently obliged here as elsewhere, to affix the label 'genuinely existent object' to the correlate of his judgement's validity, to what it judges about (LI, Inv.I, p.230-231, §31).

Once again we see the influence of Lotze as prominent.[72] The paradigmatic examples mentioned above are, in Husserl's text, drawn from

---

which makes both more explicit. See Husserl's: *The Idea of Phenomenology*, "Philosophy as Rigorous Science", and the *Ideas I* for an outline of his post-*Investigations* transcendental agenda.

[72] In his *Logik*, Lotze distinguished between real existing objects, ideal existing objects and ideal non-existing objects. In Lotze's metaphysics there are "events",

mathematics. Let us say something here about this model of concepts as 'universal objects' as this conception runs throughout the Logical Investigations and is an integral part of Husserl's theory of science in that same work influencing both his theory of a purely theoretical logic and the subsequent concept of ideational or eidetic abstraction of his (at this stage) analytic phenomenology.

To explore the notion of ideal objects as essences, following Husserl, we shall take our examples from the field of mathematics. This doctrine, i.e. of numbers and mathematical propositions as 'ideal objects' or essences, is present throughout the Investigations and grows out of certain texts of Husserl's early writings. It is, however, as presented by Husserl in the Investigations, unclear and here we shall proceed to criticise and see what consequences this theory has on other aspects of the epistemology in the Investigations. In the Prolegomena Husserl provides a brief account of the origin of mathematical knowledge very similar to that given in his first book the Philosophy of Arithmetic (1891). In Prolegomena, paragraph forty-six, for example, Husserl uses the analogy of the research domain of pure logic as being similar to that of mathematics because both are ideal domains. Like logic, no one, claims Husserl (building on points from his critique of psychologism), would regard the theories of pure mathematics as direct parts or branches of psychology. However, this being true, still: "we should have no numbers without counting, no sums without addition, no products without multiplication etc." (LI, p.109, § 46). It is only through reflection upon actual acts of arithmetical operation, writes Husserl, "that we can 'show' what a total, sum, product etc., is" (ibid). While counting and arithmetical operations viewed in this way may fall under the domain of psychology (the empirical science of mental facts in general) arithmetic does not. The domain of the natural numbers, which is the domain of research operated on within arithmetic, is said by Husserl to be exhaustively determined by the natural numbers themselves which he writes are 'ideal species'. Numbers are not 'individual facts' somehow existing in the world; they are not 'temporally

---

which occur in nature, "things", which are said to 'exist' and finally "thoughts" which are said to be 'valid' (or merely subsist in contrast to 'things'). Thoughts do not, on this view, owe their reality to either events or things (which are said to mutually interact with each other). According to Lotze, validity is a character of things that obtain [*bestehen*] but do not exist. The laws of logic for Husserl would be said to have a similar status; i.e., they would be ideal but non-existing. Husserl's pure logic however is also influenced by the logical theories of Bernard Bolzano specifically his doctrine of propositions-in-themselves ['*Sätze an sich*']. See Husserl's "A reply to a critic of my Logical Investigations" in Mohanty, J.N. (Editor), *Readings on Edmund Husserl's Logical Investigations*, Martinus Nijhoff, (The Hague, 1977), pp. 36-37 for details.

definite' (Husserl's main criterion for worldly existence). They also cannot be reduced to the presentations in which they are given. To count out 5 objects is not to create the number five within my act of counting. In the act of counting '5' is "a possible object of acts of presentation" but in itself the number '5', like all numbers, is "the ideal species of a form whose concrete instances are found in what becomes objective in certain acts of counting, within the collective whole that these constitute" (idem). In the philosophy of mathematics the position that individual numbers exist as unchanging ideal entities, highly abstract however capable of being expressed in symbols and marks but not reducible to their empirical tokens, is called Platonism. However, this is not exactly the view Husserl is taking as regards the ontological status of numbers.

Ideal unchanging 'objects' (perhaps existing in a third realm) would certainly be viewed as a psychological hypostatizing of the objects of mathematics on Husserl's view. For Husserl, the view of numbers as 'objects' somehow akin to physical objects and yet completely transcendent of the temporally determined real world would mean that an epistemic link between them and a conscious subject would be incredibly difficult, if not impossible, to establish. Rather, the viewpoint that Husserl is taking seems to be closer to a position within the philosophy of mathematics known as structuralism.[73] Like the Plationists, the structuralists also want to maintain that numbers are ontologically independent of human beliefs and of our acts of knowing them (i.e. realism in ontology) and simultaneously that they have an objective truth-value. Unlike a Platonist however, a structuralist would claim that while we (as knowers) can grasp unchanging mathematical truths, these truths are not the same as the ideal, immutable, abstract entities from which they are abstracted. On a structuralist philosophy of mathematics we know mathematical truths because our minds can grasp the patterns within the structures which underlie the mathematical knowledge. Furthermore, this knowledge is not established via any kind of empirical induction or

[73] Which in, Husserl's case, could be viewed as a kind of logical, rather than metaphysical, Platonism. See Robert Hanna's "Logical Cognition: Husserl's Prolegomena and the Truth in Psychologism", in *Philosophy and Phenomenological Research*, Vol. LIII, No.2, June 1993, pp.251-275. Proponents of structuralism among contemporary philosophers include: Paul Benacerraf, Geoffrey Hellman, Michael Resnik, and Stuart Shapiro. These thinkers do not all share one univocal conception of structuralism however and the fundamental differences pertaining to their view points are too involved to articulate here in any detail (and would furthermore take us off topic). For the sake of thoroughness it should be pointed out that Husserl's views seem most compatible with what is called 'ante-rem structuralism' i.e. the viewpoint that mathematical structures exist independently of their instantiations in particular theorems. See Shapiro (2000), pp. 263-270.

psychological 'causal' connection. In a likewise manner mathematical 'objects' for Husserl may be understood as both transcendent and as 'invariants' to be grasped intuitively within consciousness through intentional psychic acts. As Richard Tieszen notes, "Husserl argues that we…find invariants or regularities in our mathematical experience."[74] These invariants may be interpreted as a one over a many, as types of a token and, in fact, are so interpreted in many passages of the Logical Investigations. Furthermore, the invariant patterns seem to correspond to what are defined as the 'essences' of numbers by Husserl. Thus, we are told that: "Arithmetic[s]…domain of research …is completely and exhaustively determined by the familiar series of ideal species 1,2,3…In this sphere there can be no talk of individual facts" (LI: Prol. p.109, § 46). That is because, as we have noted, on Husserl's view, the individual numbers are taken to be 'ideal species'. Here we may pause and ask: "Does Husserl's view not represent a confusion between type and token?"

The number '5', for example, is an abstraction with no essence other than is found in its being a symbol. To say otherwise would be to claim that there is an ontological independence of some sort among the natural numbers which, as was stated above, would amount to some form of ontological Platonism. On a consistent structuralist view however, the 'essence' of a natural number is said to consist only in its relations to other natural numbers. As Stuart Shapiro writes: "The subject-matter of arithmetic is a single abstract structure, the pattern common to any infinite collection of objects that has a successor relation, a unique initial object, and satisfies the induction principle" (Shapiro, 2000, p.258). Yet, as we have noted, Husserl's viewpoint seems to assign a unique essence to every individual number in the structure of the natural numbers. If we view a system as a collection of objects with certain relations among them and take the abstract form of the system to be its 'pattern' or 'structure' (subsequently making this pattern or structure recognizable through acts of abstraction) we will arrive at a position very close to Husserl's. The structuralist, however, against Husserl, holds that (pure) mathematics consists in the deductive study of the structures as such (ibid, p.259) or in what could be called the 'internal relations' of the structures under consideration (Resnik, 1981, pp.539-40). This view would make notions of 'object' and 'identity' unequivocal but thoroughly relative. Also, the a priori axioms of arithmetic or geometry on this view are not established explicitly through intuitive insights into their grounds but rather given an implicit and functional definition by having at least one possible

---

[74] See Richard Tieszen "Mathematics" in the *Cambridge Companion to Husserl*, edited by Barry & David W. Smith, (Cambridge, 1995), Tieszen however, does not link Husserl's views with structuralism or any other extant philosophy of mathematics.

structure or system satisfy the axioms. The underlying logic of axiomatization is also not unique but determined by satisfiability (of the axioms) within the set-theoretic hierarchy. Thus, in the end, our structuralist position stands in a stark contrast as regards some very central points to Husserl's conception of pure logic in the Investigations. In Husserl's defence, he was quite aware of the difficulty of accounting for the form of existence of mathematical objects. As we have seen, this was the reason why he claimed that numbers and propositions etc. existed only "as mere bearers of ideal properties independent of real being or non-being" (LI, Prol: p.16, § 5). These bearers, as we have claimed, are instantiated in temporal acts of knowledge. This in turn is the essence of the 'species model' of knowledge that Husserl inherited from previous thinkers such as Lotze and Bolzano. It is note worthy that Husserl himself later takes issue with this model. In the posthumously published Experience and Judgement for example, Husserl writes that:

The irreality of objectivities of understanding must not be confused with generic universality. Since, in particular, any number of affirmative acts, of no matter how many subjects, affirm this one and the same proposition, it can have one and the same sense, and it is a great temptation to think that the proposition belongs to the various acts by which it is the sense by virtue of its generic universality…

On the contrary, now we are told that:

> Certainly, the proposition is general insofar as it refers to an infinite number of propositional acts in which it is precisely what is intended; but it is not general in the sense of generic universality, i.e., the generality of an "extension," which belongs specifically to the generality of a species, to a kind or genus, and at the lowest level, to a concrete quiddity; it is, therefore, not general in the manner of essences, which correspond to so-called general concepts such as color, tone, and the like (Husserl, 1972:1973, pp.261-62).

Husserl now seems to claim that it is a mistake to view the invariant meaning-structure of the species as somehow 'extended' thus giving it an ontological status it does not have. Later, Husserl will work out an articulation of the distinction between types of objectivities and, in particular, what he calls the broad significance between the real and irreal. To a large extent these later theories rely on Husserl's notion of the transcendental subject and its constitutive role in constituting all meaning out of what will be called 'transcendental subjectivity', however these notions-as we have seen- all have their origins within the preparatory work of the Logical Investigations.

Conclusion

In the Logical Investigations, as we have seen, there are two main approaches undertaken by Husserl in his project of grounding scientific knowledge. How these two approaches towards objective or scientific knowledge are to be reconciled is the main problem facing Husserl in that text. The first approach places the foundations of knowledge within the structure of the objective meanings of the connections of truth represented by the 'species' of propositions of knowledge. This view has been criticised above and we can summarize our criticism with the thought that, if, as seems plausible, deduction is from truth to truth and mental or conscious acts, on the contrary, can be accounted for empirically and phenomenologically then the meeting point between the ideal (axiomatically grounded) system of propositions and the insight into logical and scientific truth given through analysis of intentional acts seems a difficult one to bridge. Husserl's Logical Investigations, however, proves a very good starting point in handling the above problem, providing us in effect with one of the best sustained analysis of the interaction between subjective cognitive acts and their possible relation to the objective laws of logic.

Bibliography of Works Consulted and Cited

Bernet, Rudolf. "Husserl's theory of signs Revisited" in Edmund Husserl and the Phenomenological Tradition: Essays in Phenomenology, edited by Robert Sokolowski, (Catholic University of America Press, 1988).

———. Kern, Iso & Marbach, Eduard. An Introduction to Husserlian Phenomenology (Northwestern, 1993).

Bolzano, Bernard. Theory of Science: Attempt at a Detailed and in the main Novel Exposition of Logic with Constant Attention to Earlier Authors, translated & abridged with an Introduction by Rolf George, (Oxford, Blackwell,1972).

Brentano, Franz. Psychologie vom empirishen Standpunkte. Edited by Oskar Kraus (1874). Translated as Psychology from an Empirical Standpoint, by A. C. Rancurello, D.B. Terrell, and L. McAlister. Latest edition with an introduction by P. Simons, (Routledge, London, 1995).

Delacampagne, Christian. A History of Philosophy in the Twentieth Century, translated by M.B. Debevoise (Johns Hopkins, 1999).

Goldstern, Martin and Judah, Haim. The Incompleteness Phenomenon: A New Course in Mathematical Logic (AK Peters Natick, Massachusetts, 1995).

Hanna, Robert. “Logical Cognition: Husserl’s Prolegomena and the Truth in Psychologism”, in Philosophy and Phenomenological Research, Vol. LIII, No.2, June 1993, pp.251-275.

Heidegger, Martin. Logik: Die Frage Nach Der Wahrheit (Vittorio Klostermann, Frankfurt, 1976).

Husserl, Edmund. Collected Works Vol. V. Early Writings in the Philosophy of Logic and Mathematics, trans. by Dallas Willard, (Kluwer Academic Publishers: Dordrecht, 1994).

———. Experience and Judgement: Investigations in a Genealogy of Logic, trans. by J.S. Churchill & K. Ameriks, (Northwestern U. Press, Evanston, Ill, 1973).

———. Ideas Pertaining to a Pure Phenomenology and to a Phenomenological Philosophy: First Book, trans. by F. Kersten, (Kluwer Academic Publishers, Dordrect, 1982).

———. Collected Works VIII. The Idea of Phenomenology (Dordrect: Kluwer, 1999).

———. Logical Investigations, 2 volumes, translated by J.N. Findlay, latest edition with a new preface by Michael Dummet and Introduction by Dermot Moran. (Routledge, 2000).

———. Collected Works Vol X. Philosophy of Arithmetic: Psychological and Logical Investigations with supplementary texts from 1887-1901, trans. by Dallas Willard, (Kluwer Academic Publishers, Dordrect, 2003).

Kusch, Martin. Psychologism: A Case Study in the Sociology of Knowledge. (Routledge, 1995).

Lotze, Hermann. Logik (System der Philosophie I), Edited by Georg Misch. (Leipzig, Felix Meiner, 1912) Translated as Logic in three Books by Bernard Bosanquet (Clarendon Press, Oxford, 1888).

Mclandri, Enzo. “Thc ‘Analogia Entis’ according to Franz Brentano: A Speculative-Grammatical Analysis of Aristotle’s Metaphysics” Topoi 6 (1987), 51-58.

Miller, J. Philip. Numbers in Presence and Absence: A Study of Husserl’s Philosophy of Mathematics, (Martinus Nijhoff Publishers, The Hague, 1982).

Mohanty, J.N. (Editor), Readings on Edmund Husserl’s Logical Investigations, (Martinus Nijhoff, The Hague, 1977).

Moran, Dermot. “Husserl’s Critique of Brentano in the Logical Investigations” Manuscrito XXIII, pp.163-205, (October. 2000).

Moran, Dermot. Introduction to Phenomenology, Routledge, (London, 2000b).

Nagel, Ernest and Newman, James. Gödel’s Proof (New York University Press, 1958).

Peckhaus, Volker. "19th Century Logic between Philosophy and Mathematics"The Bulletin of Symbolic Logic, Vol. 5, No. 4. (Dec., 1999), pp. 433-450.

Philipse, Herman. "Psychologism and The Prescriptive Function of Logic" Grazer Philosophische Studien, Vol. 29, 1987, pp.13-33.

———. "The Concept of Intentionality: Husserl's development from the Brentano Period to the Logical Investigations" in Philosophy Research Archives: Vol. 12, (March 1987b), pp.293-328.

———. "Transcendental Idealism" in the Cambridge Companion to Husserl, edited by D.W & B. Smith, Cambridge University Press, (1995), pp. pp.239-322.

Paolabanchetti- Robino, Marina. "Husserl's Theory of Language as Calculus Ratiocinator" Synthese, 112, pp. 303-321, (1997).

Resnik, Michael. "Mathematics as a Science of Patterns: Ontology and Reference," Noûs Vol. 15, no. 4 (Nov. 1981); pp. 529-550.

Rollinger, Robin. Husserl's Position in the School of Brentano, Phaenomenologica series 150, (Kluwer, Dordrecht, 1999).

Shapiro, Stuart. Thinking about Mathematics: The Philosophy of Mathematics, (Oxford U. Press, 2000).

Tieszen, Richard. "Mathematics" in the Cambridge Companion to Husserl, edited by Barry & David W. Smith, (Cambridge, 1995).

Thomas. Evan Eduard. Lotze's Theory of Reality, (Longmans, Green, and Co., London, 1921).

Welton, Donn. The Other Husserl: The Horizons of Transcendental Phenomenology (Indiana U. Press, 2000).

Zahavi & Stjernfelt (Editors). One Hundred Years of Phenomenology: Husserl's Logical Investigations Revisited. Kluwer Academic Publishers (Dordrecht, 2002).

CHAPTER SIX: ELENA IRRERA

CHOOSING ONE'S OWN LIFE. INDIVIDUAL HAPPINESS AND POLITICAL EXPERTISE IN ARISTOTLE

In Book I of the Nicomachean Ethics Aristotle raises a question of great relevance for the development of his ethical and political theory: what is the best life for a man to lead? Such a question will not be answered unless we turn to contemplate the issues of human good and happiness. A modern reader might be tempted to treat the Aristotelian concept of happiness exclusively in terms of individual wishes, ambitions and preferences; but Aristotle is not adopting any "apolitical" approach to the issue of happiness, and his concern for political expertise and its role in the education of individuals with a view to a happy life is not to be downplayed.

The concept of happiness as set up by Aristotle has been a matter of vigourous debate among scholars. Besides the much controverted problem of a supposed continuity between Book I and Book X (where perfect eudaimonia is identified with theoretical activity)[75], the general lack of a unilateral approach to the various aspects which shape this notion has engendered in the past a plethora of interpretative difficulties and continues to do so in the present, which has probably been the source of a wide-and dismaying, perhaps-variety of readings. In Book I the highest good is described as the object sought by political expertise (NE I, 1094a25-28); such an expertise sets out which of the other expertises are needed for the well-

---

[75] Among the scholars who supported the view of continuity, see S. Broadie, 1991, pp. 370-373; J..M. Cooper, 1987 (where he undertakes a revision of the position assumed in J.M. Cooper, 1975, [see below]); D. Devereux, 1981; T.B. Eriksen, 1976, pp. 135-136; R.A. Gauthier and J.Y. Jolif, 1970, vol. II, pp. 891-896; W.F.R. Hardie, 1980, p. 364; R. Heinaman, 1988, p. 32; A. Kenny, 1992, pp. 29-31 (although at p. 36 he detects a lack of continuity between Books I and X with reference to the notion of self-sufficiency); D. Keyt, 1983; R. Kraut, 1989, pp. 4-5; A.O. Rorty, 1980; J.O. Urmson, 1980, pp. 14-15; J. Whiting, 1986. Conversely, doubts about the consistency of Book X with Book I have been lodged by J.L. Ackrill, 1980; A.W.H. Adkins, 1978; D.J. Allan, 1970, p. 139; J.M. Cooper, 1975, pp. 156-164; W. Jaeger, 1948, pp. 439-440; J. Moline, 1983; J.D. Monan, 1980, p. 114; T. Nagel, 1980, p.7; M.C. Nussbaum, 1986, pp. 373-377; W.D. Ross, 1923, pp. 233-234; N.P. White, 1981, pp. 242-243; and K.V. Wilkes, 1980, pp. 351-352.

being of the polis (NE I, 1094a28-b2) and, in Aristotle's view, the task of lawgivers who possess such an expertise is to foster the exercise of excellence among the citizens (NE I, 1102a7-10) with a view to both their personal happiness and the happiness of the whole of the community. On the other hand, happiness seems to be a matter of individual choice, since different people will have different beliefs as to what a happy life is; vulgar people, for instance, might suppose happiness to be pleasure or honour (NE I, 1095a22-3), whereas people of quality might go for a life of excellence, e.g. a life devoted to political activity or to activity of reflection (NE I, 1095b18-19). Again, the human good can be identified by reference to the ergon of human beings (NE I, 1097b22-1098a20), i.e. a typifying activity in virtue of which men are distinguished by the rest of living creatures; such an ergon is identified with activity of reason, and to anthrōpinon agathon will be achieved when a human being performs his own ergon well, i.e. according to some kind of excellence. Last, but not least, in Book X Aristotle engages in a persuasive apology of theoretical activity as the one allowing human beings the best and most perfect life[76], being activity performed according to the divinest part of the human soul in us[77]; a claim which, taken at a face value, would end up ruling out any possibility of achieving happiness for those who wish for a different kind of life.

The aim of this paper is to investigate the Aristotelian notion of eudaimonia in the Nicomachean Ethics by examining the role of political expertise in the actualisation of individual happiness, and the extent to which the highest good pursued by expert lawgivers for the well-being of the political community is compatible both with individual choice as to how to live one's own life and with the idea of a human ergon. In the first section of the present work I will try to provide evidences for the idea that the highest good which expert lawgivers seek to actualise in the polis is mainly of an ethical kind; a good which, as I propose to demonstrate, will be perfect and complete only when grounded in and supported by theoretical activity. In section II I will suggest that happiness, in Aristotle's thought, is not a systematic composite of different ends, but involves the possibility of choosing between two different kinds of excellence for the sake of a happy life: the ethical and the intellectual. I intend to demonstrate that, notwithstanting the alleged superiority and higher perfection of intellectual activity, men can choose either or both as desirable in themselves.

---

[76] See *NE* X, 1177a12-1178a8.
[77] See *NE* X, 1177a13-17.

The Highest Good as The Object of Political Expertise

Aristotle's concern for the role played by political expertise in the pursuit of human happiness is expressed in Book I of the Nicomachean Ethics. He begins by claiming that

'Every sort of expert knowledge and every inquiry, and similarly every action and undertaking, seems to seek some good. Because of that, people are right to affirm that the good is "that which all things seek"'[78].

What is immediately worth stressing here is that the good is described in terms of ends. The idea of aiming emerging through this passage is not concerned with individual, psychological inclinations; by contrast - unlike people of different kind, who tend to some specific goods on the basis of various motives and with a view to the satisfaction of their personal wishes and ambitions - each craft, activity, practice and project seems to seek its own good as the only telos that makes a specific thing the thing it is and reveals its nature, or as the end the achievement of which constitutes a good performance of a specific pursuit[79].

The impersonal character of crafts, activities, undertakings etc. is confirmed in the following lines. As Aristotle goes on, there is a certain difference among ends: some are activities, whereas others are products over and above the activities themselves, by their nature better than the activities

[78] *NE* I, 1094a1-2. Tr. Rowe, 2002, from which all the passages of the *Nicomachean Ethics* mentioned in this article will be taken. In this introductory passage Aristotle has been charged with a serious fallacy: that of moving from 'all chains must stop somewhere' to 'there is somewhere where all chains must stop'. See J.L. Ackrill, 1980, pp. 25-26; G.E.M. Anscombe, 1963, p. 34; P.T. Geach, 1958, p. 112. For an illuminating comment on the passage see S. Broadie, 1991, p. 8: "According to one not uncommon interpretation of the opening lines of the *Nicomachean Ethics*, we must start by lowering our expectations of Aristotle as logical thinker. It is charged that in his anxiety to justify the view that there is a supreme good for human beings, he begins by committing a gross fallacy, this being the move [...] from 'Each thing aims at some good' to 'There is some good at which each thing aims'". Unlike the ommentators mentioned above, Broadie acquits Aristotle of the fallacy, by pointing out that Aristotle's move from 'every craft etc. aims at some good' to 'the good is that at which all things aim', although not formally valid, does not pretend to formal validity, because the second sentence is still in need of argument (p. 9). As she explains at p. 10, when Aristotle mentions 'the good to which all things aim', he is not pointing to the specific end proper to each practical pursuit and corresponding to its peculiar nature; rather, he seems to be referring to "a universal end "beyond" the specific or partial ones", or, in other words, "a practical point of convergence in relation to all activities and projects". For other commentators who acquitted Aristotle of the fallacy see B.A.O. Williams, 1962.

[79] See S. Broadie, 1991, p. 16.

from which they spring (NE I, 1094a3-6). Goods are not necessarily related to each other; for instance, health, which is the end of medicine, will not have anything to do with the art of shipbuilding and its products, i.e. ships, nor will victory, which is the end of generalship, relate in any way to the end of household management; however, there are cases in which some activities fall under some single capacity: the craft of bridle-making, for instance, falls under horsemanship insofar as bridles are tools employed in the latter form of expertise, and horsemanship, in its turn, falls under generalship and will be employed in the art of war (NE I, 1094a8-14).

However, those crafts which fall under others are not taken as having the same value as the crafts under which they fall. As Aristotle says,

> 'In all activities the ends of the controlling ones are more desirable than the ends under them, because it is for the sake of the former that the latter too are pursued'[80].

At this stage of the argument we are still unable to explain why some things are more desirable than others, nor are we entitled to provide a possible reason of their being worth pursuing. Only in the following lines will Aristotle show that the goodness of both whatever is pursued for the sake of an higher end and the higher ends themselves is determined by and grounded in the goodness of a further end, representing the ultimate reason of their being undertaken. As we might notice in the passage at issue, Aristotle is significantly trying to reduce the multiplicity of the ends, which seems to pave the way for the idea of a unique and supreme good for the sake of which everything else is pursued. However, the existence of such a good is not implied in what it has been said so far. In order to be the cause of the goodness of things pursued for the sake of it, it should be the condition of the goodness of the other things and not depend on any further end[81].

These requirements are satisfied by the rest of the argument, where the highest good is hypothesized as something wished for because of itself and, at the same time, the cause of our wishing the other things:

> 'If then there is some end in our practical projects[82] we wish for because of itself, while wishing for the other things we wish for because of it, and we do not choose everything because of something else (for if that is the case, the sequence

[80] *NE* I, 1094a14-16.

[81] See S. Broadie, 1991, p. 11: "we are not yet entitled to conclude to a supreme good. A 'further good' would be supreme only if (1) it underpins the goodness of every one of the specific ends; (2) it is unique in having this function (3) there is nothing on which it in turn depends for its value as others depend on it".

[82] This is Rowe's translation for 'τέλος ἐστὶ τῶν πρακτῶν'.

> will go on to infinity, making our desire empty and vain), it is clear that this will be the good, i.e. the chief good'[83].

This passage implies that we cannot speak of the goodness proper to those expertises, acvitities, choices etc. that are pursued for the sake of a higher end unless we hypothesize a final end as the point to which all of them must halt; without such an end, every pursuit would be 'empty' and 'vain'[84]. We might expect Aristotle to say expressely what is the nature of the supreme good; but what Aristotle confines himself to doing is to reveal the expertise to which it belongs: the political. At NE I, 1094a26-b2 he introduces such an expertise, i.e. the one to which the highest good belongs, by stressing its 'architectonic' character:

> 'It [the highest good] would seem to belong to the most sovereign, i.e. the most 'architectonic' [expertise][85]. Political expertise[86] appears to be like this, for it is this expertise that sets out which of the expertises there needs to be in cities, and what sorts of expertise each group of people should learn, and up to what point'.

The highest good seems to be 'architectonic' in relation to the other goods in the same way as a controlling activity is 'architectonic' in relation to the others; however, it is not a controlling activity of whatever kind, nor does it control a restricted sphere of expertises. Political expertise employs all the other expertises needed in cities and, presumably, includes them in the organisation of the political community as components of the goods at which it aims. The highest good is not realised by means of expertises which, once the good has been achieved, are dismissed; on the contrary, they are permanently employed in political life - as we might read Aristotle's claim that the architectonic expertise 'makes use of the practical expertises that remain'[87] - so that they will turn out to be a vital part of the organisation of a political community. Accordingly, the end of the most 'architectonic'

---

[83] *NE* I, 1094a18-22.

[84] See S. Broadie, 1991, p. 13: "My desire must settle at some point because until it does I do not actually desire anything for the sake of which I then desire the things through which I can accomplish it. If I shift along the series discarding each as an ultimate objective in favour of the next, then not only do I form no desire for a particular objective, but my general desire *for good* (whatever the good may be) is 'empty and futile' - not because it cannot be implemented to the maximum, but because it cannot be implemented *at all* unless I fix on some given objective as good, and set about acting for the sake of it".

[85] δόξειε δ' ἂν τῆς κυριωτάτης καὶ μάλιστα ἀρχιτεκτονικῆς.

[86] ἡ πολιτικη

[87] *NE* I, 1094b4-5. Rowe's reading retains 'πρακτικαῖς' at 1094b4.

expertise, i.e. 'the human good'[88], will contain those of the rest. That the highest good belonging to the political expertise contains the ends of the goods pursued for its sake seems to show that it is a comprehensive good. Goods like health, for instance, might be desirable in themselves and therefore be regarded as complete goods; but they will not include goods of other kinds, like victory, which is the end of generalship. The highest good, by contrast, is thought to embrace all the different ends pursued by human beings.

Aristotle is also keen on stressing two features belonging to the highest good: completeness and self-sufficiency. As for completeness, at NE I, 1097a28-30 he speaks of a good which is more complete than others, and so most complete:

> 'if there is some one thing alone that is complete, this will be what we are looking for, and if there are more such things than one, the most complete of these'.

Completeness is introduced in relation to what is worth pursuing for itself and not because of something else, given that, as explains in the following lines,

> 'what is worth pursuing for itself is more complete than what is worth pursuing because of something else, and what is never desirable[89] because of something else is more complete than those things that are desirable both for themselves and because of it; while what is complete without qualification[90] is what is always desirable in itself and never because of something else'[91].

The idea of completeness can be also detected in the second attribute attached to the highest good: self-sufficiency, which makes life 'desirable and lacking in nothing'[92].

So far, Aristole has tried to provide some characteristic features of the highest good, without making clear what the highest good actually is. One possible suggestion might be to read Book I of the Nicomachean Ethics in the light of Book A of the Metaphysics, where Aristotle introduces the notion of sophia in relation to the highest good; understanding the highest good embodied by sophia might be useful towards understanding how the notion

---

88 *NE* I, 1094b7: 'τἀνθρώπινον ἀγαθόν'.

89 αἱρετόν.

90 ἁπλῶς τέλειον.

91 *NE* I, 1097a30-34.

92 See *NE* I, 1097b14-15: 'τὸ δ' αὔταρκες τίθεμεν ὃ μονούμενον αἱρετὸν ποιεῖ τὸν βίον καὶ μηδενὸς ἐνδεᾶ'.

of highest good is employed in the Ethics. It is interesting that in Book A sophia is described as the most divine and honourable among the various forms of epistēmai[93], and also as a good desirable because of itself, given that people do not seek after it for the sake of something else[94]. Again, at Met. A, 982b4 sophia is described as archikōtatē de tōn epistēmōn and, similarly to political expertise in the Nicomachean Ethics (which is described as an architectonic form of expertise[95], is regarded as ruling over the other sciences that depend on it, given that it knows the end for the sake of which everything is done in nature[96]. As Aristotle goes on to say, the end of each thing is the good, and, in nature as a whole, the end is the highest good[97].

The highest good in which sophia seems to consist is surely intellectual, given that at Met. A, 982b7-10 sophia is said to speculate on the first principles and the causes. If so, should we suppose that the highest good pursued by sophia is the same as that pursued by political expertise? It is surely easy to notice some similarity between the Nicomachean Ethics and that of the Metaphysics, especially in their start, given that at Met. A, 980a21 it is claimed that every men aims at knowledge by nature[98], and the idea of 'tending' that emerges here is apparently similar to that expressed at NE I, 1094a1, where it is said that every sort of expertise tends to some good[99]; it might be supposed that such a similarity reveals that Aristotle intends to speak of the highest good in the same terms in both works.

However, it seems to me that, unlike in the Metaphysics, in the first Book of the Nicomachean Ethics Aristotle's main interest is not the highest good conceived as a search of causes; as a matter of fact, in the Metaphysics nowhere does Aristotle say that the highest good is sought after by political expertise, nor in the passages of the Nicomachean Ethics analysed so far does he mention sophia with reference to the highest good. It seems that the exercise of political expertise involves the employment of the other kinds of expertises needed for the well-functioning of the polis, among which are generalship, household management, rhetoric[100], differently from intellectual

---

93 See *Met.* A, 983a5, where Aristotle describes *sophia* as 'θειοτάτη καὶ τιμιωτάτη'.

94 See *Met.* A, 982b24-28: 'δῆλον οὖν ὡς δι' οὐδεμίαν αὐτὴν ζητοῦμεν χρείαν ἑτέραν [...] μόνη γὰρ αὕτη αὑτῆς ἕνεκέν ἐστιν'.

95 *NE* I, 1094a26-28.

96 *Met.* A, 982b5-6.

97 *Met.* A, 982b6-7.

98 'Πάντες ἄνθρωποι τοῦ εἰδέναι ὀρέγονται φύσει'.

99 The verb employed in the *Ethics* to indicate the activity of 'tending to something' is 'ἐφίεσθαι'.

100 These kinds of expertises are mentioned at *NE* I, 1094b3.

activity, which, rather, might presuppose the knowledge of disciplines like mathematics, astronomy and geometry.

As the whole of the Nicomachean Ethics shows, ethical excellence and practical thought seem to be the main focus of Aristotle's investigation; that he devotes so much space to issues like ethical excellence, wisdom, justice and friendship reveals that the aspect of the highest good he has in mind is mainly of ethical nature. Although a virtuous state might promote intellectual activity as an important ingredient of its happiness[101], intellectual activity per se will not determine the self-sufficiency of the community. More specific evidence in support of the idea that the highest good pursued by political expertise is of an ethical kind is that at NE I, 1102a8-9 Aristotle suggests that that the aim of the true political expert is excellence:

> 'the true political expert will have worked at excellence more than anything'.

Although Aristotle does not openly say which kind of excellence the lawgiver is working on, the kind of excellence Aristotle seems to have in mind here is ethical, given that what the political expert wants is

> 'to make the members of the citizen-body good, and obedient to the laws'[102].

Given that laws are established with a view to the promotion of justice in the polis, it seems that the highest good pursued by political experts is mainly ethical. Again, a passage which seems to confirm such a belief may be found at Pol. III, 1280b7-9, where it is said that

> 'any polis which is truly so called, and is not merely one in name, must devote itself to the end of encouraging goodness[103]. Otherwise, a political association sinks into a mere alliance, which only differs in space from other forms of alliance where the members live at a distance from one another'[104].

Although Aristotle does not specify the kind of aretē for which the state should have concern, it seems clear that the virtue in question is ethical; it would be unlikely that lack of intellectual excellence would make a community an alliance rather than a true state. By contrast, lack of ethical

---

[101] See *Met.* A, 981b23-25, where it is said that mathematical sciences were founded for the first time in Aegypt, thanks to the leisure bestowed to the priestly class in the country.

[102] *NE* I, 1102a9-10: τοὺς πολίτας ἀγαθοὺς ποιεῖν καὶ τῶν νόμων ὑπηκόους.

[103] δεῖ περὶ ἀρετῆς ἐπιμελὲς.

[104] Tr. Barker, 1948.

excellence and justice might cause fellow-citizens to lack reciprocal well-wishing and trust, as though they did not belong to the same community. Although this passage is not directly connected to the context of human happiness, still it seems to show the main concern of true political experts and, as we know from Book I of the Nicomachean Ethics, happiness is pursued by political expertise.

Having suggested that the end of political expertise is mainly of an ethical kind, how shall we understand Aristotle's appeal to desirability with reference to the highest good? How might the idea of the desirability of the highest good be reconciled with the object of a political expertise that supervises the exercise of the other activities? I will now deal with happiness under the perspective of individual choice, and then take up the importance of political expertise as the source of the highest good for the community.

Happiness and Individual Wishes

At NE I, 1095a20 ff. Aristotle introduces various views held by people about the highest good sought by political expertise. Both ordinary people and people of quality identify the highest good with happiness, but there is a general disagreement on the nature of happiness. Ordinary people usually identify happiness with things like pleasure, wealth or honour, and pick out many other things depending on particular individuals and circumstances; for instance, when a man falls ill, he will seek after health, whereas a poor man will wish for wealth as a source of happiness. By presenting such a variety of conceptions of happiness, he shows his awareness that people actually choose how to be happy. Their choice, however, will depend on the quality of people. The most vulgar individuals suppose happiness to be pleasure, which is the reason why they favour the life of consumption (NE I, 1095b14-17); besides that, two other outstanding kinds of life are mentioned: the political life and the life of reflection (NE I, 1095b17-19). In this section of Book I, the idea of preference and choice seems to be more prominent. For instance, vulgar people

> 'decide in favour of a life that belongs to grazing cattle'[105],

> whereas

> 'those who tend towards a life of action, go for honour'[106].

[105] *NE* I, 1095b19-20.
[106] *NE* I, 1095b22-23.

As Aristotle explains, the values pursued by such people are superficial in that they are not complete and self-sufficient. Honour, for instance, seems to be located in those doing the honouring rather than in the person who receives it, whereas what is really good is something that belongs to a person (NE I, 1095b24-26). As he says in the following lines, excellence is of greater value than any unspecified honour (NE I, 1095b30-32). I believe that, when at NE I, 1095b23 Aristotle claims that honour is pretty much the end of political life, he is probably referring to a common belief or, in other words, to the reason why most people decide to engage in political activity, but not to his personal view of political life as a kind of life based on the exercise of ethical excellence. What Aristotle is looking for, which is at the same time what political expertise should look for, is a good worth desiring, i.e. a good which is the cause of the goodness of the other things pursued by human beings; in that case, a good like wealth, for instance, will never be the highest good: the life of the money-maker is chosen under compulsion of need, and wealth is looked for, for the sake of something else (NE I, 1096a5-7).

When Aristotle describes the highest good as the end of political expertise, he does not seem to regard it as an exclusive possession of political experts, given that he devotes wide attention also to what people generally wish for[107]; however, given that the highest good must be the same for each individual and for the polis as a whole[108], it seems that individual choice of the highest good will not be arbitrary, in that there might be individual conceptions of happiness which do not accord well with the well-being of the state. In that case, if we want to preserve the view that people might identify their personal happiness with the highest good, we must assume that the end of political expertise should at least be part of their good, if not the whole of it.

It might be thought that, in Aristotle's view, the function of political expertise is simply to protect the individual in his personal pursuit of the good, whatever this good is; however, this does not seem to be a fundamental concept in his argument on the highest good[109], given that he is not saying that such an expertise confines itself to providing the conditions for the individual realisation of the highest good; rather, Aristotle claims that the highest good is the end itself of political expertise, and such a good is produced through the well-functioning of the polis by means of the other skills involved in the political organisation. This implies that, as Hardie puts it,

---

[107] Cf. G. Santas, 2001, pp. 224-225, who calls '*orectic*' the good seen under the perspective of human desires.

[108] See *Pol*. VII, 1234a5-7.

[109] See S. Broadie, 1991, pp. 16-17

> 'the thinking required for a man's arrangement of his own life is of the same kind, although on a smaller scale, as the thinking of the statesman'[110].

If we take Hardie's suggestion as plausible, as I do, to see what the main value for the political expert is will enable us to understand what the good for any individual is.

As we have already seen in the passage at NE I, 1102a9-10, Aristotle claims that the task of the true political expert is to work at excellence so as make the citizens good and obedient to the laws, and, as I proposed, such a kind of goodness is an 'ethical' one. Aristotle is not simply saying that some degree of ethical activity is required of citizens for the well-functioning of the whole polis; rather, what he intends to show is that happiness is realised only when the citizens are made good individuals through the laws, presumably laws which prescribe a virtuous behaviour. If the happiness of the community is the same as the happiness of its members taken individually, exercise of ethical excellence turns out to be an indispensable aspect of human happiness. This conclusion, as I will try to show in the next section of this paper, does not seem to be incompatible with the idea that theoretical activity allows the best and the most complete life; nor - as I think - does it conflict with the thought that there might be different individual preferences as to what the highest good should be. That ethical activity is an indispensable ingredient of human happiness does not necessarily entail that it is its only ingredient. Some people, for example, might go for intellectual activity as intrinsicably preferable to ethical activity, provided that the pursuit of it is compatible with the ethical good of the polis or, at least, does not undermine it. As I am going to argue in the following section, even when activity according to ethical excellence is not regarded by everyone as the highest good in all its completeness, ethical good should always be part of human eudaimonia, and it might even be desirable in itself, without this preventing intellectual activity from being pursued as an intrinsically desirable good.

Indeterminacy of the Highest Good and The Function Argument

We have seen how Aristotle, after describing the highest good as the end of political expertise, shifts his attention to an idea of happiness which seems to be more connected to human wishes and choices. Under this perspective, the highest good might be viewed as the end in the light of which human

[110] W.F.R. Hardie, 1980, p. 17.

choices can be explained and valued[111]. In this section I will try to show that the completeness and self-sufficiency proper to the highest good are not a matter of including all the other goods, but, rather, of intrinsic desirability and, to some extent, of individual choice. On my reading of Aristotle, the highest good is in a way something indeterminate.

A passage which might be indicative of some indeterminacy is NE I, 1097b1-6, where Aristotle claims:

> 'this [happiness] we do always choose because of itself and never because of something else, while as for honour, and pleasure, and intelligence, and every excellence, we do choose them because of themselves (since if nothing resulted from them, we would still choose each of them), but we also choose them for the sake of happiness, supposing that we shall be happy through them. But happiness no one chooses for the sake of these things, nor in general because of something else'.

As I read this passage, Aristotle here mentions honour, pleasure, intelligence and the other excellences only as possible examples of things with which people may identify their own happiness, and not as fixed ingredients of it. The idea of a certain kind of indeterminacy might also be supported by NE I, 1095a20-26 and 1095b14-23, where - as we have already seen - Aristotle illustrates different kinds of life and of dominant values varying according to the quality of people. But such beliefs he intends to dismantle, with a view to showing that the highest good - differently from what people of a certain quality might believe - is activity of reason according to some kind of excellence, even if we admit, as I do, that - in Aristotle's view- individuals of whatever kind aim at happiness[112].

---

[111] See A. Kenny, 1992, p. 3: "To explain a choice, an agent will have to specify the good pursued. In doing so, he will show how is action is related, directly or indirectly, to an ultimate end. Asked to give a reason for an action, an agent may either explain that the action is a means to some end, or say that it is valued for its own sake". Cf. R. Kraut, 1989, p. 201: "To determine which goods are for the sake of which others, we must go beyond purely causal questions about which goods produce which others; we must ask what the appropriate norms are for regulating our activities"; and p. 208: "The only way to make decisions about the relative weights of various ends is to use that highest good as the target of all our actions. So it is not merely the best of all goods - it is also the standard by which we fix the value of all else".

[112] Whether *eudaimonia* is the end for whose sake all action *is* undertaken (which is called by J. McDowell 'indicative thesis'; see J. McDowell, 1980, p. 1) or that for whose sake all action *ought* to be undertaken ('gerundive thesis') is a question still open to debate. See for instance A. Kenny, 1977, who supports the gerundive thesis, vs J. McDowell. See also J. McDowell, 1980, p. 2: "Aristotle himself has a specific view about what kind of life constitutes *eudaimonia*. He certainly does not hold that everyone aims to lead *that* kind of life. But this yields no argument against attributing

However, when I claim that Aristotle's concept of happiness is in a way indeterminate, I am not referring to the idea that people of any kind whatever might have different conceptions of happiness. Rather, the indeterminacy to which I would like to point has to do with the man who is good by nature, and with his possibility of choosing between a life in which either the ethical/political or the theoretical activity plays a dominant role. I hope my point will emerge more clearly in the last part of this section, where I will try to reconcile the idea that the highest good is mainly ethical with the Aristotelian view that the best life is activity in accordance with practical reason.

Let us now take up NE I, 1097b1-6, and examine another difficulty. It is interesting that the goods pursued for the sake of the highest one do not seem to be mere means to its achievement, since they are chosen at the same time because of themselves. It might be wondered how some things are pursued both because of themselves and for the sake of a higher good. In my view, this is possible if and only if such goods are components of the highest good; in other words, the goods desired for the sake of the highest one would not be merely instrumental to it, but they would contribute to shaping it, in the same way as the activities pursued for the sake of political expertise are not instrumental to it but are included as parts of political life and determine its organisation.

The absence of a list of specific goods belonging to human happiness has puzzled scholars, who have proposed different solutions to the problem; however, two main lines of thought might be identified: the 'inclusivist' and the 'dominant'[113]. As Ackrill explains, those who embrace the first tendency assume that the Aristotelian highest good includes a compound of values and activities of different kind, as he himself does in thinking that eudaimonia necessarily includes every valuable activity[114]; on the face of it, those who support the idea of a dominant end suppose the highest good to consist of just one valuable activity or good: either intellectual[115] or practical activity[116]. I

---

to him a thesis like (i) [i.e. the gerundive]". At p. 1, with reference to *NE* I, 1102a2-3 (where it is claimed that "it is for the sake of happiness that we all do everything else we do"), he suggests that "there seems to be no prospect of taking this to express a gerundive thesis".

113 The terminology which distinguishes 'inclusive ends' from 'dominant ends' was introduced by W.F.R. Hardie, 1965, p. 291.

114 See J.L. Ackrill, 1980, p. 22; cf. D. Devereux, 1981; T.H. Irwin, 1985 and 1986; M.C. Nussbaum, 1982; A.W. Price, 1980.

115 Among the representatives of the 'intellectualist' view see A.W.H. Adkins, 1978; J.M. Cooper, 1975 (but he will support a form of inclusivism compatible with the idea of a dominant end in J.M. Cooper, 1987); W.F.R. Hardie, 1965 (but he will slightly revise his theory in W.F.R. Hardie, 1979, p. 42, where he points out that to acknowledge the 'dominance' of a single end is not incompatible with an inclusive

believe that both views on human happiness are in a way incorrect. On the one hand, were we to follow the inclusivist view as Ackrill conceives it, we would probably do violence to the Aristotelian view of happiness. Nowhere in the Nicomachean Ethics does Aristotle seem to list the valuable goods included within happiness[117]; moreover, happiness would be achieved only in the case that all the goods it supposedly includes are pursued over an entire life.

A lucid refutation of Ackrill's arguments is provided by Kraut, who, in his work Aristotle on the Human Good criticizes his belief that

> 'happiness should not be identified with any single good such as ethical activity, or contemplation, or honor, or pleasure. For any such good is just one among many, and no matter how desirable it may be, it is always less desirable than the combination of that good and some other'[118].

As Kraut points out, those commentators who, like Ackrill, support an inclusivist view, try to justify their position by appeal to the attributes of completeness and self-sufficiency attached to the highest good in NE I, 1097b14-20, where happiness is described as choiceworthy and in need of nothing[119]. But if we assumed that happiness is something desirable in itself and lacking in nothing, even the lack of one good would make its achievement impossible for any individual. Even more, we would not be able to explain the reason why Aristotle resorts to the notion of a human ergon in order to explain what the human good is, and his praise of the superiority of contemplative life with reference to the concept of eudaimonia in Book X would turn out to be pointless.

The 'dominant' view, on the other hand, generally precludes the possibility of taking more than a single good or activity as desirable in itself. For instance, were happiness determined by a single, prominent ingredient like contemplation, we should exclude ethical activity as an essential component of happiness. Again, according to the dominant view, the so-called 'external' goods (e.g. health, wealth or good luck), which might be

---

view); R. Heinaman, 1988; A. Kenny, 1977 (cf. A. Kenny, 1992, where he attempts to argue that in the *Nicomachean Ethics* Aristotle proposes a dominant view of happiness, whereas in the *Eudemian Ethics* an inclusive view); D. Keyt, 1978; R. Kraut, 1989; G.R. Lear, 2004; T. Nagel, 1972.

[116] See J.O. Urmson, 1988.

[117] See S. White, 1990, quoted by A. Kenny, 1992, p. 28.

[118] R. Kraut, 1989, p. 267.

[119] See R. Kraut, 1989, p. 267. See also A. Kenny, 1992, p. 24. According to the majority of supporters of the 'dominant' view, the notions of completeness and self-sufficiency contribute to show the inherent perfection of intellectual activity.

taken as ingredients of a happy life (although not in the same way as virtuous activity)[120], would surely be ruled out from the notion of eudaimonia.

As I believe, the so-called 'function argument', (whose role in Aristotle's theory of happiness has been the subject of broad disagreement among commentators[121]) turns out to be crucial in our attempt to find out what the highest good really is. At NE I, 1097b25-8 Aristotle expresses his intention to investigate happiness as a human good; in his view, the definition of happiness will come about only once the function of human beings has been established. By 'function' (ergon) he seems to mean a typifying activity, based on some kind of expertise or excellence, which expresses the nature of particular individuals or things; he mentions as examples of specific functions those of the flute-player, the sculptor and any expert, but also the function of each part of the body[122]. In the same way, he is looking for the specific function that belongs to human beings qua human beings; as he says at NE I, 1097b25-33:

> 'just as for a flute-player, or a sculptor, or any expert, and generally for all those who have some characteristic function or activity, the good-their doing well-seems to reside in their function, so too it would seem to be for the human being, if indeed there is some function that belongs to him. So does a carpenter or a shoemaker have certain functions and activities, while a human being has

[120] The role played by external goods in the pursuit of a happy life in Aristotle's thought is not clear, and still open to dispute. Supporters of the inclusivist view hold them to be components of a happy life (see for instance J.L. Ackrill, 1980; J.M. Cooper, 1987; T.H. Irwin, 1985; and M.C. Nussbaum, 1986. Conversely, those commentators who subscribe to a 'dominant' view of happiness take external goods as conditions of, rather than ingredients of happiness. A different position is held by J. Annas in Sherman (ed.), 1999. Annas maintains that Aristotle fails to reconcile the role played by external goods in the pursuit of happiness with his own theory of *eudaimonia*, whose core is represented by virtuous activity. My personal view is that, in Aristotle's theory, external goods are necessary conditions of, but not the core of happiness. See for instance the account of happiness provided in the *Eudemian Ethics*. At *EE* I, 1214b14-17, where it is claimed that, just as being healthy is not the same as the things without which it is not possible to be healthy, so living well is not the same as the things without which living well is impossible.

[121] A list of commentators who tried to analyse this argument is provided by G. Santas, 2001, at p. 236. Santas mentions J.L. Austin, 1967, and J.M. Cooper, 1975, who believe that Aristotle engages in an unnecessary and too abstract argument, whereas T.H. Irwin, 1985, and W.F.R. Hardie, 1980, are presented as supporters of the view that the function-argument is devised by Aristotle just to express 'his own view' of happiness; finally, G. Anagnostopoulos, 1980, and D. Keyt, 1983, maintain that the function argument is a central piece in Book I. The same position, to which I subscribe, is maintained by Santas, 2001, and R. Kraut, 1989.

[122] See *NE* I, 1097b25-26; 1097b30-1.

> none, and is by nature a do-nothing? Or just as an eye, a hand, a foor, and generally each and every part of the body appears as having some function, in the same way would one posit a characteristic function for a human being too, alongside all of these?'

The human function will not consist either in taking in nutriment and growing, which is typical of plants, or in a kind of life of perception, which is proper to animals[123]. The sort of life that remains as characteristic of human beings is the life of what possesses reason; of this, as he says, one element 'possesses reason' insofar as it obeys reason, whereas the other actually has it[124].

The Aristotelian argument so presented raises some essential problems: (i) how can men have functions? (ii) Are they designed for a purpose? (iii) And, even if they did, how would this coincide with what is good for them[125]? A Reasonable answer to the first two questions is provided by Santas, who maintains that the idea of a human function is indebted to Aristotle's biological and psychological investigations. On Santas' view, the human function is not to be understood in terms of roles and occupations but, rather, in terms of characteristics, capacities or potentialities of an object; for example, the notion of function is applied in Aristotle's biology to organs of animals without ever making any appeal to a designer of animals or their parts[126]. When it comes to human beings, the notion of 'ergon', expressing some kind of functionality, is connected to Aristotle's concern for human activity, and not for excellence as a mere possession. As he points out at NE I, 1095b31-1096a2, when he speaks of the different kinds of life chosen by different people,

> 'excellence [too] appears somewhat incomplete: for it seems to be possible actually to be asleep while having one's excellence, or to spend one's life in inactivity, and furthermore to suffer, and to meet with the greatest misfortunes; and no one would call the person who lived this kind of life happy, unless to defend a debating position'.

Aristotle intends to move from the notion of 'human ergon' to that of 'human good'; what he is going to show in the remainder of his argument (which will enable us to answer the third question formulated above) is that the human good will be achieved only when the distinctive function of human beings is practised well. As I believe, the idea of a human ergon is a necessary, but not a sufficient condition for the understanding of what the

---

[123] See *NE* I, 1097b33-1098a3. Cf. *De An.* I, ch. 1; *De An.* II, ch. 4; *De An.* III, ch. 9.

[124] See *NE* I, 1098a3-5.

[125] Such questions are raised by J. Whiting, 1988, at p. 33.

[126] See G. Santas, 2001, p. 241.

human good is[127]. As Aristotle points out at NE I, 1098a7-8, the function of a human being is

> 'activity of soul in accordance with reason, or not apart from reason'[128],

and only at NE I, 1098a16-18 it is claimed that

> 'the human good turns out to be activity of soul in accordance with excellence (and if there are more excellences than one, in accordance with the best and the most complete)'[129].

It seems that, when Aristotle claims that the human function is activity according to excellence, he is not confining himself to investigating a feature that makes mankind different from the rest of living creatures; rather, he seems to be looking for the highest degree of excellence achievable by men just by virtue of their human nature, while being conscious that many individuals are engaged in inferior kinds of occupation. However, such a definition does not provide any further details as to what kind of excellence should be pursued in order to live a happy life, nor does it explain what Aristotle means by 'completeness' when he says that happiest will be the life according to the most complete excellence. In the next section I will try to provide an answer to these questions.

Two candidates for Human Happiness: Theoretical and Ethical Excellence

Two kinds of excellence - in Aristotle's view - might be regarded as mostly desirable in themselves and self-sufficient: the intellectual and the ethical; both kinds of excellence seem to accord with the human function. As he explains at NE I, 1103a3-7, these kinds stem from two different rational aspects that are proper to human soul: a part which participates in reason and another which possesses reason fully.

---

[127] For a similar view see A. Gomez-Lobo, 1989; at p. 171 of his article he criticizes Wilkes' view that the *ergon* of a man has itself the criterion for determining the value of actions and/or persons (cf. K.V. Wilkes, 1980, p. 343). Conversely, at p. 176 Gomez-Lobo maintains that "in order to identify a good *F* (or the goodness for *Fs*) two logically distinguishable steps must be taken: first, the *ergon* of *Fs* has to be ascertained in a non-evaluative manner, and then, an evaluative judgement as to what counts as a good performance of that *ergon* has to be passed. This latter judgement is equivalent to finding the *arete* or *aretai* corresponding specifically to the class of *Fs*. The *ergon* and the good are not strictly identical".

[128] ἐνέργεια κατὰ λόγον ἢ μὴ ἄνευ λόγου.

[129] τὸ ἀνθρώπινον ἀγαθὸν ψυχῆς ἐνέργεια γίνεται κατ° ἀρετήν, εἰ δὲ πλείους αἱ ἀρεταί, κατὰ τὴν ἀρίστην καὶ τελειοτάτην.

As for theoretical excellence, that it might be identified with human happiness is explicitly said at NE X, 1177a12-17, where it is also described as an excellence of the highest kind:

> 'But if happiness is activity in accordance with excellence, it is reasonable that it should be activity in accordance with the highest kind[130] and this will be the excellence of what is best. Whether, then, this is intelligence or something else, this element that is thought naturally to rule and guide, and to possess awareness of fine things and divine ones, - whether being, itself too, something divine, or the divinest of the things in us, it is the activity of this, in accordance with its own proper excellence, that will be complete happiness'.

It is interesting that, when he introduces theoretical excellence, he explicitly refers to what was said earlier (προειρημένα)[131], as though he meant to establish some continuity between Book I and X of the Nicomachean Ethics; that this is his real intention is confirmed by the following lines, where - once again - he says that happiness is not a disposition but a kind of activity, and that it is desirable in itself and self-sufficient. As Aristotle explains at NE X, 1177a17-22, theoretical activity is the highest because intelligence too is the highest of human things, and its objects are the highest knowables; moreover, it is taken as the most continuous, probably because, more than practical activities, it involves a minimal physical effort[132].

It might be concluded, then, that Aristotle believes the most complete happiness to be activity accordance with intellectual, and not with ethical activity, which in Book X is deemed to bring about only the 'second happiest' life[133]. As he explains, just things and all the other kinds of things we do according to various ethical excellences appear to be human, not divine, and many of them require the body rather than the highest part of the soul; moreover, ethical excellence seems to be bound up with the affective states rather than with pure nous[134]. Again, theoretical excellence seems to be more self-sufficient than ethical activity, since the latter involves decisions or doings, as is required by a fully human life[135], whereas the former might be practised in isolation and does not need the cooperation of other human beings.

---

130 κατὰ τὴν κρατίστην.

131 *NE* X, 1176a32-3.

132 Cf. Broadie's commentary of the passage, in S. Broadie-C. Rowe, 2002, p. 442.

133 See *NE* X, 1178a9: Δευτέρως δ' ὁ κατὰ τὴν ἄλλην ἀρετήν.

134 See *NE* X, 1178a10-16.

135 See *NE* X, 1178a34-b7.

Shall we conclude, then, that Aristotle takes intellectual activity as the only, dominant end in human life? According to my reading, that intellectual excellence is taken as the best and the most complete, and the one capable of guaranteeing the best life, does not allow us to claim that he does. After all, at the beginning of his account of eudaimonia he presents his inquiry on the highest good as 'political'[136], and, as we have seen, the highest good is the object of political expertise, the aim of those who possess such an art being to make the citizens ethically good and obedient to the laws. Again, there seem to be some ingredients of happiness, e.g. friendship[137], which are strictly related to ethical goodness and do not have to do with contemplation.

To the alleged superiority of intellectual to ethical activity one might object that, although in Book X theoretical activity is claimed to be more self-sufficient than the practical, since the person engaged in reflection does not need the things necessary to carry an action through, even in practical activity a degree of self-sufficiency will be provided; political expertise, which sets out which of the other expertises are needed in the polis, might make life of each member of the community self-sufficient through a fair distribution of functions according to each individual's competence, and to this self-sufficiency also activity of citizens according to justice will contribute. In other words, although individuals by themselves are not self-sufficient, still their belonging to the political community will provide them with the necessary sources for their pursuit of ethical excellence. In that case, ethical excellence, just like the intellectual, might be constitutive of human happiness, although in a different way from the theoretical.

As for political activity, which I take as a form of 'activity according to ethical excellence', that it is the activity to which the highest good belongs shows that it has to do with the most complete among practical goods; moreover, that at NE I, 1094a26-7 it is defined as the most sovereign expertise, i.e. the most 'architectonic'[138], might lead us to think that it falls under the notion of 'most complete activity' introduced in the definition of human happiness. That political activity is more complete than other activities is confirmed at NE I, 1094b8-10, where Aristotle says that to achieve the good for a whole city is 'finer and more godlike' than doing it for one's own good.

Likewise, even without being political experts, common people might practise an ethically virtuous life and live according to the most complete excellence of character: justice. That justice is regarded by Aristotle as a

[136] See *NE* I, 1094b10-11.

[137] See *NE* VIII, 1155a5-6, where Aristotle says that nobody would choose to live without friends.

[138] κυριωτάτη καὶ μάλιστα ἀρχιτεκτονική.

more complete excellence than others is confirmed at Book V of the Nicomachean Ethics, where he suggests that it is

> 'complete excellence, only not without qualification but in relation to another person'[139],

and, as he goes on to say at NE V, 1129b30-3, justice is regarded as the mightiest of excellences:

> 'it is complete excellence to the highest degree because it is the activation[140] of complete excellence; complete, too, because the person who possesses it has the capacity to put his excellence to use in relation to another person as well, and not just by himself'.

As we see, Aristotle describes justice as a disposition capable of 'activating' the rest of the ethical excellences, as he confirms at NE V, 1129b29-30 when he quotes the saying 'justice gathers in excellence entire'. In this respect, activity according to justice will be finer more complete than activity according to just a few of the ethical excellence it includes.

If so, what is happiness? Should it be identified with a life whose dominant character is the intellectual, or with a life devoted to ethical/political activity? Or, finally, does it consist of a life in which both kinds of excellence are practised as intrinsic goods? I believe that the solution to this problem lies in the indeterminate character attached by Aristotle to the notion of happiness: it seems to be a good which - to some extent - involves human choice. That in Book I neither the contemplative nor the practical excellence are explicitly introduced as constitutive ingredients of happiness - besides being due to the provisional character of the account of the highest good at the beginning of the Nicomachean Ethics - might entail that a margin of freedom is given to us as to which life is desirable to lead because of itself. Some people, for instance, might go for a life in which ethical excellence is the cause of their doing all the rest; they might express their human nature through a kind of non-theoretical excellence and lead a happy life in this way. By contrast, other people might take a life of contemplation as preferable to a life based on the exercise of justice or of political excellence, and in this respect they might do everything for the sake of it. They might perform their work in the polis well mainly with a view to getting the necessary resources for a good life and devote themselves to the activity of reflection, or they might behave justly with a view to avoiding conflict with other people and so attaining the harmony required by their favourite activity compromised. In a

---

[139] See *NE* V, 1129b25-27.

[140] χρῆσις.

different case, they might behave according to ethical virtue as an inherently worthwhile good, although taking activity of contemplation as having a higher value.

But a question is still left unsolved: what does Aristotle mean when he suggests that, if there is an excellence which is better and more complete than others, this will be the excellence according to which one ought act with a view to one's own happiness? I would suggest that - on his view - the 'best and the most complete' activity is the intellectual. This seems to be confirmed in Book X of the Nicomachean Ethics (see NE X, 1177a12-22), where he explains that, if eudaimonia is activity in accordance with excellence, it should be activity in accordance with the highest kind, i.e. the theoretical. Those people who identify their happiness with a life devoted exclusively to ethical/political activity will attain only a 'second-best' life, provided that such an activity does not allow the pursuit of contemplation. When at NE X, 1178a9 Aristotle claims that 'second happiest' is the life according to ethical excellence, I take him to mean that such a life is less complete than a life in which contemplative activity not only does not exclude, but even involves activity in accordance with ethical excellence. In fact, it might be hypothesized that contemplative activity prompts a wider awareness of the need to act justly; the importance of activity according to ethical excellence in human life, for instance, might be understood more fully only after having grasped the essence of human nature and contemplated the position that human beings occupy in the cosmos[141]. I maintain that, when in the Nicomachean Ethics Aristotle speaks of a person engaged in theoretical activity and not committed to an ethical life as the most valuable kind of life, he does not seem to be thinking of an individual isolated from the world outside; theoretical excellence must be practised within a political context. In other words, contemplative activity does not exclude the application of ethical values in one's life, just insofar as a person pursuing such an activity is a member of a political community, even when ethical activity is only a restricted part of one's happiness, or something which is not desirable in itself.

However, it is worth noticing that the function-argument, devised by Aristotle with a view to explaining what the human good is, does not per se suggest that happiness is confined to activity according to intellectual excellence[142]. This might imply that both ethical and intellectual excellence might be regarded as intrinsically good and complete ingredients of eudaimonia, although only intellectual activity will ensure the best and most complete happiness. As I claim, according to the Aristotelian theory of

[141] See J. Lear, 1988, p. 8: "The project of understanding the world lies at the bottom of who we are".

[142] See T.D. Roche, 1988, p. 183, and A. Kenny, 1992, p. 29.

happiness, an individual might choose a less-perfect kind of life than the theoretical, i.e. the ethical, and still be happy, although less happy than a person who places more weight on the exercise of intellectual excellence. This assumption seems to accord well with Aristotle's view that the highest good is the object of political expertise, and that such a good is mainly ethical. The idea that we can choose which excellence is desirable towards happiness does not entail that an entire life is to be lived in accordance with an exclusive pattern; in fact, nowhere does Aristotle state that one kind of life should be assumed as precluding another one. To act according to aretē in a complete life, then, does not necessarily amount to acting exclusively according to a single, most complete excellence. As it seems, in Aristotle's view the nature of happiness as the highest good is not supposed to be something already settled.

References

J.L. Ackrill, 'Aristotle on Eudaimonia', in Rorty (ed.), Essays on Aristotle's Ethics, Berkeley, University of California Press, 1980.

A.W.H. Adkins, 'Theoria versus Praxis in the Nicomachean Ethics and the Republic', Classical Philology 73(1978), pp. 297-312.

D.J. Allan, The Philosophy of Aristotle (2d ed.), New York, Oxford University Press, 1970 (first published in London, Oxford University Press, 1952).

G. Anagnostopoulos, 'Aristotle on Function and the Attributive Nature of Good', in Depew (ed.), The Greeks and the Good Life, Indianapolis, Hackett, 1980.

G.E.M. Anscombe, Intention (2d ed.), Ithaca (N.Y.), Cornell University Press, 1963 (first published in 1957, Oxford, Blackwell).

J.L. Austin, 'Agathon and Eudaimonia in the Ethics of Aristotle', in Moravcsik (ed.), Aristotle, A Collection of Critical Essays, New York, Anchor Books-Doubleday, 1967.

E. Barker, Aristotle. Politics. Translated with an Introduction, notes and Appendixes, Oxford, Oxford University Press, 1948, (first published 1946).

S. Broadie-C.J. Rowe (eds.), Aristotle. Nicomachean Ethics. Translation, Introduction, and Commentary, Oxford, Oxford University Press, 2002.

S. Broadie, Ethics with Aristotle, Oxford, Oxford University Press, 1991.

J.M. Cooper, Reason and Human Good in Aristotle, Cambridge, (Massachusetts), Harvard University Press, 1975.

J.M. Cooper, 'Contemplation and Happiness: A Reconsideration', Synthese 72(1987), pp. 187-216.

D. Devereux, 'Aristotle on the Essence of Happiness', in O'Meara (ed.), Studies in Aristotle, Washington, Catholic University of America Press, 1981, pp. 247-270.

T.B. Eriksen, Bios Theoretikos: Notes on Aristotle's Ethica Nicomachea, X, 6-8, Oslo, Universitetsforlaget, 1976.

R.A. Gauthier and J.Y. Jolif, L' Éthique à Nicomaque. Introd., Traduction et commentaire (2d ed.), Louvain, Publications Universitaires, 1970 (first published 1958-1959).

P.T. Geach, 'History of a Fallacy', in P.T. Geach, Logic Matters, Berkley (CA), University of California Press, 1958, pp. 1-13.

A. Gomez-Lobo, 'The Ergon Inference', Phronesis 34(1989), pp. 170-184.

W.F.R. Hardie, 'The Final Good in Aristotle's Ethics', Philosophy 40(1965), pp. 277-295.

W.F.R. Hardie, 'Aristotle on the Best Life for a Man', Philosophy 54(1979), pp. 35-50.

W.F.R. Hardie, Aristotle's Ethical Theory, Oxford, Clarendon Press, 1980 (2d. ed.; first published 1968).

R. Heinaman, 'Eudaimonia and Self-sufficiency, in the Nicomachean Ethics', Phronesis 33(1988), pp. 31-53.

T.H. Irwin, 'Permanent Happiness: Aristotle and Solon', in Annas (ed.), Oxford Studies in Ancient Philosophy III, Oxford, Oxford University Press, 1985, pp. 89-124.

T.H. Irwin, 'Stoic and Aristotelian Conceptions of Happiness', in Schofield-Striker (eds.), The Norms of Nature, Cambridge, Cambridge University Press, 1986, pp. 205-244.

W. Jaeger, Aristotle: Fundamentals of the History of His Development (2d ed.), London, Oxford University Press, 1948, (first published 1923).

A. Kenny, 'Aristotle on Happiness', in Barnes-Schofield-Sorabji (eds.), Articles on Aristotle. 2. Ethics and Politics, London, Duckworth, 1977, pp. 25-32.

A. Kenny, 1992, Aristotle on the Perfect Life, Oxford, Clarendon Press, 1992.

D. Keyt, 'Intellectualism in Aristotle', Paideia, Special Aristotle Issue, 1978, pp. 138-157; repr. in Anton-Preus (eds.), Essays in Ancient Greek Philosophy, vol. 2, Albany, State University of New York Press, 1983, pp. 364-387.

R. Kraut, Aristotle on the Human Good, Princeton, Princeton University Press, 1989.

G.R. Lear, 2004, Happy Lives and the Highest Good: An Essay on Aristotle's Nicomachean Ethics, Princeton, Princeton University Press, 2004.

J. Lear, Aristotle: The Desire to Understand, Cambridge, Cambridge University Press, 1988.

J. McDowell, 'The Role of Eudaimonia, in Aristotle's Ethics', Proceedings of the African Classical Associations 15(1980), pp. 1-15.

J. Moline, 'Contemplation and the Human Good', Nous 17(1983), pp. 37-53.

J.D. Monan, Moral Knowledge and Its Methodology in Aristotle, Oxford, Clarendon Press, 1980.

T. Nagel, 'Aristotle on Eudaimonia', Phronesis 17(1972), pp. 252-9; repr in Rorty (ed.), Essays on Aristotle's Ethics, Berkeley, University of California Press, 1980, pp. 7-14.

M.C. Nussbaum, Aristotle, in T. James Luce (ed.), Ancient Writers Greece and Rome, Volume I: Homer to Caesar, New York, Charles Scribner's Sons, 1982, pp. 404-410.

M.C. Nussbaum, The Fragility of Goodness, Cambridge, Cambridge University Press, 1986.

A.W. Price, 'Aristotle's Ethical Holism', Mind 89(1980), pp. 338-352.

T.D. Roche, 'Ergon and Eudaimonia in Nicomachean Ethics I: Reconsidering the Intellectualist Interpretation', Journal of the History of Philosophy 26(1988), pp. 175-194.

A.O. Rorty, Akrasia and Pleasure: Nicomachean Ethics Book 7, in Rorty (ed.), Essays on Aristotle's Ethics, Berkeley, University of California Press, 1980.

W.D. Ross, Aristotle, London, Methuen, 1923.

G. Santas, Goodness and Justice. Plato, Aristotle and the Moderns, Oxford, Blackwell, 2001.

J.O. Urmson, 'Aristotle's Doctrine of the Mean' in Rorty (ed.), Essays on Aristotle's Ethics, Berkeley, University of California Press, 1980, pp. 157-170.

J.O. Urmson, Aristotle's Ethics, Oxford, Basil Blackwell, 1988.

N.P. White, 'Goodness and Human Aims in Aristotle's Ethics', in O'Meara (ed.), Studies in Aristotle, Washington, Catholic University of America Press, 1981, pp. 225-246.

J.E. Whiting, 'Human Nature and Intellectualism in Aristotle', Archiv für Geschichte der Philosophie 68(1986), pp. 70-95.

J.E. Whiting, 'Aristotle's Function Argument: A Defense', Ancient Philosophy 8(1988), pp. 33-48.

B.A.O. Williams, 'Aristotle on the Good – A Formal Sketch', Philosophical Quarterly 12(1962), pp. 289-296.

K.V. Wilkes, 'The Good Man and the Good for Man', in Rorty (ed.), Essays on Aristotle's Ethics, Berkeley, University of California Press, 1980, pp. 341-357.

CHAPTER SEVEN: NOELLE RUMBALL

DANGEROUS LIAISONS: A SOLUTION TO THE GETTIER PROBLEM

Traditional epistemology gives too much weight to the intuitions that actual knowledge requires knowledge in a space of nearby possible worlds, and that there is only one way to have knowledge – the strongest kind. But I will offer a view that rejects these particular intuitions, and I will do it through denying that certain Gettier cases really are counterexamples to the JTB analysis. I will then offer a different definition of knowledge, based around a distinction between types of Gettier cases provided by Stephen Hetherington: Helpful and Dangerous cases. A Helpful Gettier case is one in which there is some odd fact that is instrumental to the subject's belief or justification that p, for example, the dog in sheep's clothing. A Dangerous Gettier case is one in which there is some hidden danger that would topple the subject's belief or justification that p, but as it happens does not actually do so, for example, Henry and barn-façade land. On my view, only the Dangerous Gettier cases are knowledge, while the Helpful Gettier cases are not. Dangerous cases have the advantage that they seem inherently the same as paradigm cases of knowledge; the difference is not between the cases themselves, but in the circumstances surrounding them. It is through the rejection of the above intuitions that makes such a definition of knowledge possible.

My definition and non-standard intuitions have several good points. For one, we get a lot more of the everyday knowledge denied by other variations on the JTB analysis. However, it does not commit to knowledge just being true belief. Particularly, because my theory supports the intuition that there can be better or worse ways of knowing just as there are better or worse ways of being justified, it has the result that the JTB analysis does not come apart in a way that has inspired some philosophers to give up on ever solving the Gettier problem.

In this paper, I will be discussing an account of propositional knowledge, using a non-standard approach to the debate. In order to properly discuss how my proposal is non-standard, I will first introduce how I will use the standard terms of the debate. Then, I will discuss the standard intuitions, and suggest some reasons for doubting those intuitions. I will follow by outlining Stephen Hetherington's distinction between types of Gettier cases[143] in a move

[143] This distinction is from Hetherington 2001.

towards my non-standard account of knowledge, which I will then define. I will begin by introducing the language of the debate.

Until 1963, the traditional, or at least default, analysis of knowledge was as follows:

> JTB: An epistemic subject S knows a proposition p if, and only if, S has a justified[144], true, belief that p.

In 1963, Edmund Gettier introduced two counterexamples ('Gettier cases') to this definition of knowledge, and so introduced the 'Gettier problem': there are cases in which JTB is satisfied, but there is some Strange Occurrence[145] affecting the situation, which, standard intuitions tell us, stops S from knowing that p[146].

The first responses to the Gettier problem were to suppose that a justified true belief is necessary, but not sufficient for knowledge, and so the approach was to propose some fourth condition: a JTB + X theory – for example, JTB with the added stipulation that S cannot have any false evidence in his justification[147]. However, epistemologists soon became experts at creating new Gettier cases for each new proposed definition, and no fourth condition has ever been widely accepted.

Another approach was to deny that a justified true belief is even necessary for knowledge, and to try to find some completely different way of defining

---

144 In this paper, I shall treat justification as neutral to the internalism/externalism debate, for reasons of space.

145 This is Hetherington's term, from Hetherington (2001). See section 3.2 particularly.

146 For example, suppose Sally is in her office on the 50th floor of a skyscraper. Suddenly, Jim, the man from the next partition over, screams "I just can't take it anymore!" grabs his chair, smashes the window, and jumps out. However, on the 49th floor, there is a window-cleaner, and Jim falls directly onto his scaffolding. Just as Jim is gathering his wits, the window-cleaner pulls out a gun and shoots him. Sally has a justified true belief that Jim is dead (he is dead, and there is no way he could have survived a fall of 50 stories), but due to the Strange Occurrence (there was a crazy window cleaner with a gun), Sally doesn't know that Jim is dead.

147 See D.M. Armstrong *Belief, Truth and Knowledge*, especially p. 147 for support of this theory, and Richard Feldman "An Alleged Defect in Gettier Counter-Examples" for a way to paraphrase away the false evidence in *S*'s justification that *p*.

knowledge[148]. But again, new Gettier cases were proposed to cast doubt on these theories. It was beginning to look as if there could never be a theory capable of escaping the Gettier problem. However, there is one more option, not usually considered: what if all Gettier cases are not true counterexamples to JTB? Denying certain types of Gettier cases as counterexamples leaves open the possibility of finding some JTB + X theory that can work as a definition for knowledge. In this paper, I will use Hetherington's distinction between types of Gettier cases in order to propose an amendment of JTB which accepts one sort and rejects another sort of Gettier case as counterexamples. But before I do that, I must first examine the kinds of intuitions that have precluded such an approach being adopted.[149]

The standard intuitions I am interested in here are the ones that affect how Gettier cases are classified, since part of my argument is that certain Gettier cases should be admitted as cases of knowledge. I will argue against two particular intuitions: first, that actual knowledge requires counterfactual knowledge; and second, that there is only one way of propositionally knowing. I will argue that it is possible to have actual knowledge without having counterfactual knowledge, and that there are better and worse ways of knowing a proposition.

To clarify the issue, I will focus on two particular "problem" cases. The first is Henry and the barns[150]:

Henry is driving along a country road, first through 'barn land', where there are lots of real barns along the road. Then, unbeknownst to him, he passes into 'barn-façade land', where most, but not all, of the barns are clever papier mâché imitations. Henry, pointing out one of the few real barns in

---

[148] For example, in "Justified Judging", Alexander Bird suggests that knowledge is primitive, and that justification that must be explained in terms of knowledge, rather than the other way around. Reliabilist accounts of knowledge, like Nozick's counterfactual definition of knowledge (to be discussed later), also tend not to make use of justified true beliefs as necessary.

[149] Stephen Stich, in "Normativity and Epistemic Intuitions," questions whether the standard intuitions are in fact held outside of Western epistemology, and has applied psychologically-styled testing to determine the effects of socio-economic and ethnic backgrounds on people's intuitions regarding whether standard Gettier cases should or should not be classified as counterexamples to JTB. Although his essay is not very good *philosophy*, running more to (imperfect) scientific method – hypothesis, experimental results, conclusion – it serves the purpose of inspiring discussion on the issue.

[150] This example is adapted from Alvin Goldman's in "Discrimination and Perceptual Knowledge," pp. 772-3

barn-façade land, has a justified true belief that 'that is a barn'[151]. But does he know it?

The second case I shall discuss is Jill and the assassinated dictator[152]:

Jill lives in a South American country run by a dictator. One day, he is assassinated. But only one news bulletin gets out with this information before the facts are hushed up by the government. Jill hears only the first, true, news bulletin, so she has a justified true belief that the dictator is dead. (We will assume she hears the news from a generally reliable source as regards truth.) But everyone else hears only the false bulletins, and has justified but false beliefs that he has only been wounded (or perhaps some hear both bulletins, and are unsure which to believe, reserving judgement). Does Jill know the dictator was assassinated?

One standard argument as regards these "Gettier" cases runs as follows: Henry and Jill were just too lucky to have the justified true beliefs that they did, and therefore too lucky to be ascribed knowledge. If Henry had seen a fake barn, rather than a real one, he would still have made the judgement, 'that is a barn'. If Jill hadn't only heard the first, true, news bulletin, she would have falsely believed the dictator was alive. Notice that Henry and Jill's luck can be cashed out by making counterfactual claims. So, these "Gettier" cases depend on the notion that knowledge ascription in certain nearby possible worlds makes a difference to knowledge ascription in the actual world. Also, they depend on the notion that knowledge is absolute. For, if knowledge admitted of degrees, luck would not necessarily imply non-knowledge; rather it would imply lower calibre knowledge. For example, justification is traditionally better or worse depending on luck, among other things. I will discuss below why I think the above intuitions are not necessarily the right ones to adopt, beginning with the importance of counterfactual knowledge.

There are two accounts of knowledge which I will use to explore the intuition that actual knowledge requires counterfactual knowledge: Robert Nozick's subjunctive conditional account[153], and Gilbert Harman's defeasibility account[154]. I will argue that both of these accounts are untenable, and that their problems can be traced back to their appeal to this intuition. On Harman's view, a subject S knows that p if, and only if, S has a justified true

---

[151] We will assume in this and in all following cases where observation is part of the justification that lighting is good, eyesight is 20/20, the object in question is not a mere pinpoint on the horizon, etc.

[152] This example is adapted from Gilbert Harman's *Thought*, p. 75

[153] See Nozick, p. 81.

[154] See Gilbert Harman, pp. 75-80, for his account of defeasibility theory of knowledge.

belief that p, and there is no fact q, such that if S were to be aware of q, S would no longer know that p. There cannot be any facts ‘in the wings’ that would ‘defeat’ the subject’s knowledge, if he were to know about them. This rules out lucky cases like Henry and Jill, because there are facts such that, if Henry and Jill knew about them, their knowledge would be defeated. Nozick has a different view, where a subject S knows proposition p only if, (1) if it were the case that p, then S would believe that p; and (2) if it were not the case that p, then S would not believe that p. This ‘truth-tracking’ method also excludes lucky knowledge, because the counterfactuals will not hold in such cases: there are near possible worlds where Henry and Jill were not lucky enough for their beliefs to track the truth in the requisite way. So, both Harman’s and Nozick’s accounts are phrased counterfactually, and thus depend on the intuition that counterfactual knowledge is required for actual knowledge. I will now discuss the rather unpleasant implications for accepting such an account.

The first problem is the relationship between counterfactual knowledge ascription and scepticism. Accepting Nozick’s view lands you squarely in scepticism, since all of our everyday knowledge fails (2) in the case where any sceptical hypothesis is possible. Nozick’s response to this was to claim that all sceptical hypotheses describe possible worlds that are too far away to be relevant for knowledge ascription in the actual world, but that is having your cake and eating it. If one claims that counterfactual knowledge is required for actual knowledge, it just isn’t convincing to go on to say, well, that’s only true when it doesn’t land you in scepticism. When that happens, it just isn’t relevant anymore.[155] There is a similar problem with the defeasibility theory, since human beings are not omniscient. It is impossible to rule out the possibility that there is information ‘out there’ that could defeat knowledge, including our everyday knowledge[156]. Now, no one has yet

[155] If one is inclined to accept that Nozick is not helping himself to too much in his response to scepticism, consider that if there is any fortuity in how *S* acquired his knowledge that *p*, say watching a couple of seconds of a documentary while channel hopping, *S* (arguably) must be denied knowledge. For example, if it was a very badly researched or highly biased documentary, and hence wrong in its assertion that *p*, then *S* wouldn’t be able to tell by watching a few seconds of it (failing condition 2 of Nozick’s account). Any second-hand knowledge (reading books, watching documentaries) can be similarly called into question.

[156] Although this is an internalist-style objection, neither epistemic subjects nor those ascribing knowledge to them have the information required to rule out every possible defeater.

defeated scepticism, but it looks like both of these accounts of knowledge lead to it more easily than is necessary.

The other argument against requiring a counterfactual back-up for our actual knowledge ascriptions is that changing the example only slightly seems to make a difference in how it gets classified. Just imagine that Henry is only in barn-façade land for a few minutes – just long enough to make the judgement 'that is a barn' – and then he immediately turns around and goes back into barn land. The case has not actually changed, all I have done is introduced something that Henry does after he forms his justified true belief. Yet, I would find it strange to say that he does not know that 'that is a barn' in this augmented version. Similarly, imagine that Jill reads the true news bulletin on an airplane on her way back to England, and the English media doesn't ever print a falsified story about the assassination, so she keeps her belief that the dictator was assassinated for the rest of her life. Again, it seems odd to say that Jill does not know. This undermines the support for accepting Harman's or Nozick's theory, because part of the reason for accepting such theories is that they purport to ascribe knowledge correctly in every case. But, if these new examples are cases of knowledge, neither Harman's[157] nor Nozick's theory is able to accept them as such. This calls into doubt the intuition that was pushing their respective theories.

From this, it looks like counterfactual knowledge does have some affect on actual knowledge, since changing the story slightly makes a difference. However, I do not think that it makes the difference between full knowledge and no knowledge. Rather, it is a question of quality. This leads to the second intuition I wish to question, that there is only one way of knowing – the best way.

Epistemologists do not deny that 'knowledge-of' and 'knowledge-how' can be better or worse. What they do deny is that propositional knowledge ('knowledge-that') can be analysed in this way. However, this was not always strictly the case. Historically, understanding was an important factor in propositional knowledge: one could only know what was self-evident, or what was deductible from self-evident propositions. So, a better or worse understanding of why something is true could give a better or worse way of knowing. Although now most people take it for granted that we can and do gain knowledge from experts, there is a sense in which it would be nice to distinguish between being taught something and deducing it for oneself. Since, although for the most part I trust what I read in the major newspapers, I accept that journalists are all biased in one way or another, and I am always

[157] In some fairness to Harman, he only states that we are "less likely" to ascribe knowledge to Henry and Jill's (original) cases (Harman, p.75) than we are to ascribe it to paradigm cases.

prepared to consider that the paper might well be wrong. In this age of the information highway, I think it is more important than ever to have some sort of a distinction between knowing deductively from self-evident truths, and learning from the experts. However, although I think the distinction is important, I have no wish to deny that both methods are knowledge-conferring. Again, there is a difference in quality between different ways of knowing.

My claim is that you can know a fact in a better or worse way by dint of your knowledge of the collection of facts that that particular fact falls under – for example, if Henry knows all about barn-façade land (let's say he lived there as a child) then Henry's Gettier problem immediately disappears. Knowing all about his surroundings will either stop him from making judgements about barns, or, he will know from experience which are the real barns and which aren't. Regardless, his topical information changes the way in which he knows that p.[158]

Once we consider that there are better and worse ways of knowing, it is possible to use counterfactual knowledge ascriptions as a guide to which end of the gradient actual cases fall in. Accepting this leads to a way out of the Gettier problem, since justification and knowledge no longer have to come apart every time there is less than ideal justification[159]. However, this is still not a full account of knowledge. It remains to be seen whether these non-standard intuitions are anything more than will-o'-the-wisps.

If you allow that the quality of the justification affects the quality of ways of knowing, or at the very least how robust that knowledge is, then we have an analysis that better reflects the way we conceive of knowledge in practice. It accepts that counterfactually knowing does have an effect on actually knowing, yet it doesn't force us to lose our everyday knowledge. So, given that we want to support these intuitions, how does one go about giving an account of knowledge that will solve the Gettier problem? I will begin by following Stephen Hetherington's lead in dividing Gettier cases into two types: Helpful Gettier cases and Dangerous Gettier cases[160]. I will then argue that, while Helpful cases are counterexamples to JTB, Dangerous cases are

---

[158] There is an interesting parallel between this interpretation of knowledge, which has a sort of *factual* division of labour (where physicists know in a better way that atoms are made up of quarks than I do, since I'm still not sure what quarks are) and Hilary Putnam's division of *linguistic* labour in his essay "The meaning of 'meaning'."

[159] Linda Zagzebski argues that the Gettier problem is insoluble along these lines in "The Inescapability of the Gettier Problem".

[160] See section 3.2 of Hetherington 2001.

not. Finally, I will propose an account of knowledge that excludes the Helpful Gettier cases.

There is a common thread to every Gettier case: they all have some Strange Occurrence, which is what drives the wedge between justified true belief and knowledge. It is through classification of the Strange Occurrences that we get the distinction between types of Gettier cases. Looking at the various instances of Strange Occurrences, some are Helpful, where the Strange Occurrence is crucial for the epistemic subject to have a justified true belief at all. Without it, ceteris paribus, the subject would find himself in a close possible world where he does not believe that p, is not justified in believing that p, or where p is not true. An example of a Helpful Gettier case, with a Helpful Strange Occurrence, is as follows, the dog in sheep's clothing:

Say you are walking through a field with rolling hills. You see what looks like a sheep. But the sheep-like thing that you see is not actually a sheep. It's Gromit the sheepdog, in disguise. There is in fact a sheep (Shawn) in the field, and Gromit is in disguise to keep him company, yet still provide protection. You can't actually see Shawn; he's behind the hill and beyond your sight. But you do have a true belief that there is a sheep in the field, and it is justified by Gromit's sheep-like appearance, so you have a justified true belief. But do you know that there is a sheep in the field? [161]

If it weren't for the Strange Occurrence, Gromit being in disguise, you would either believe that there were no sheep in the field, or reserve judgement (noting that there are hills around you where sheep could hide). You would clearly have neither a justified true belief, nor knowledge. So, a Helpful Gettier case is one in which taking away the Strange Occurrence yields a paradigm case of non-knowledge.

A Dangerous Gettier case is one in which taking away the Strange Occurrence yields a paradigm case of knowledge. There, the Strange Occurrence is a hidden danger to the epistemic subject's justified true belief, but he still has that justified true belief in spite of the danger. Had there been no Strange Occurrence and everything else unchanged, he would have had a perfectly normal justified true belief that p, with no problems involved in satisfying the JTB analysis. Jill and the assassinated dictator and Henry and the barns are both Dangerous Gettier cases. In Jill's case, if the Dangerous

[161] This is an adaptation from an example in Roderick Chisholm *Theory of Knowing*, p.93. The adaptations are so that you do not form the false belief that what you see before you is a sheep. You only form the true belief that there is a sheep in the field. Also, Gromit being dressed as a sheep is causally 'hooked up' to there being a sheep in the field. If Shawn were somewhere else, Gromit would be as well. The awkwardness in 'fixed' cases like this will be discussed below.

Strange Occurrence of the false news report hadn't happened, she, and everyone else, would have had justified true beliefs (and knowledge) that the dictator was dead. In Henry's case, without the Dangerous Strange Occurrence that barn-façades are common to the neighbourhood, he would have had a justified true belief (and knowledge) that 'that is a barn'.

I have argued above that in the cases of Henry and Jill, there are reasons to ascribe knowledge, and the same reasons apply to other examples of Dangerous Gettier cases. But these reasons do not apply to the Helpful Gettier cases. There is no story we can tell that makes them more like cases of knowledge without drastically changing the situation; taking away a Helpful Strange Occurrence takes away knowledge as well. By definition, Dangerous cases will sound better given a different spin on the story, but Helpful cases will never have this aspect. The role that Strange Occurrences play in forming justified true beliefs ensures that.

Helpful Gettier cases cannot be accepted as cases of knowledge. In every Helpful Gettier Case, there is a Helpful Strange Occurrence which is instrumental in the epistemic subject's forming a justified true belief. So, in these cases, the subject does not have his justified true belief on his own merits – he has been helped by some weird quirk to form a belief he could not have formed on his own. This help is often in the form of false evidence, or tricks to avoid including false evidence[162]. Take the dog in sheep's clothing. Notice that, not having actually formed the false belief that Gromit is a sheep, you have no false evidence in your justification. But this means that not thinking consciously about a particular stage of your justification that p has given you knowledge that p. Using this manoeuvre is a Helpful Strange Occurrence. Without reasoning in this funny way, you ought not to have been ascribed justification for the belief that there is a sheep in the field. False justification, or justification resting on false premises, is not real justification. The epistemic subject in a Helpful Gettier Case is being given credit where no credit is due – it is only because of the Strange Occurrence that he has a justified true belief at all.

Conversely, in a Dangerous Gettier Case, the epistemic subject has formed his justified true belief on his own, and the Dangerous Strange Occurrence only serves as a possibility of failure. The subject is getting due credit for reasoning to the best of his abilities given his situation. So, Dangerous cases are cases of knowledge, not only because under certain

[162] Gettier's original examples are good for this: they included false evidence, yet are still used with modifications so that the false evidence is not consciously used in the reasoning process. See Hetherington 2001 p. 98 for his argument that an amended version of one of Gettier's examples is still knowledge, despite tricking its way out of false evidence.

descriptions it seems like they should be, but also because to deny that Dangerous cases are knowledge is to deny almost all of our everyday knowledge as well. Recall my discussion of counterfactual knowledge above. A Dangerous Gettier case just is a case in which we deny knowledge because of a lack of counterfactual knowing. Crudely, we are imagining circumstances in which the hidden danger is no longer hidden, and denying knowledge based on those circumstances. But, as discussed previously, if we depended on counterfactual knowledge for actual knowledge, then we would have very little knowledge about anything around us. Hence, in order to claim that we have knowledge in everyday situations, we are forced to accept that Dangerous Gettier cases are on a sliding scale from precarious ways of knowing, that are still fairly clear case of knowledge (like the augmented Henry and Jill cases) to extremely shaky ways of knowing[163].

So, my proposed definition of knowledge is a JTB + X theory. It takes into account that cases of actually knowing ought not to require counterfactually knowing, and that there are many ways of knowing, some good, some bad. It also acknowledges the consequence of accepting these intuitions, namely, that Dangerous Gettier cases are in fact cases of knowledge:

JTB_D: An epistemic subject S knows a proposition p in way w if, and only if, S has a justified true belief that p, and there was no Helpful Strange Occurrence playing a role in S's particular w. Where w is then better or worse depending on the extent to which it is affected by Dangerous Strange Occurrences.

There are good reasons for accepting my amended interpretation of JTB. First of all, it gives a clear answer to disagreements over whether or not Gettier cases are counterexamples to knowledge: Helpful Gettier cases are, and Dangerous Gettier cases are not. And the distinction between Helpful and Dangerous Gettier cases is equally clear-cut. My definition also gives us access to our everyday knowledge, yet leaves room for healthy doubt – you can know something while still being open to the idea that there only luckily isn't some fact you're not aware of that could take away that knowledge. Because we have accepted that there are good and bad ways of knowing, there is no longer a tension in describing it in terms of justification. As long as you can have a better justification, you can have a better way of knowing.

[163] An example of this latter sort would be the case in which Jill reads a false news bulletin minutes after she has read the true one, and we question whether or not she knew the dictator had been assassinated during those few minutes. Or, suppose Henry had continued to drive through barn-façade land for weeks, and mistakenly taken hundreds of fake barns to be real.

And finally, it supports the intuition that actually knowing does not require counterfactually knowing.

Works Cited

Armstrong, D.M. 1973. Belief, Truth and Knowledge. London: Cambridge University Press.

Bird, Alexander. "Justified Judging." (unpublished paper) accessed on November 14th 2002, and can be found at: http://homepages.ed.ac.uk/ajbird/research/papers/Justified_Judging.pdf.

Chisholm, Roderick. 1989. Theory of Knowing. 3rd edition. Englewood Cliffs: Prentice-Hall.

Feldman, Richard. 1974. "An Alleged Defect in Gettier Counter-Examples." Australasian Journal of Philosophy 52.

Gettier, Edmund. 1963. "Is Justified True Belief Knowledge?" Analysis 23.

Goldman, Alvin. 1974. "Discrimination and Perceptual Knowing." Journal of Philosophy 54.

Harman, Gilbert. 2000. "Selection from Thought." Epistemology: An Anthology. Oxford: Blackwell Publishers Inc.

Hetherington, Stephen. 2001. Good Knowledge, Bad Knowledge: On Two Dogmas of Epistemology. Oxford: Clarendon Press.

Nozick, Robert. 2000. "Knowledge and Scepticism." Epistemology: An Anthology. Oxford: Blackwell Publishers Inc.

Putnam, Hilary. 1975. "The meaning of 'meaning'." Mind Language and Reality: Philosophical Papers Volume 2. Cambridge: Cambridge University Press.

Stich, Stephen. "Normativity and Epistemic Intuitions." (forthcoming paper in Philosophical Topics) accessed on October 24th 2002, and can be found at http://ruccs.rutgers.edu/ArchiveFolder/Research%20Group/Publications/NEI/NEIPT.html

Zagzebski, Linda. 1994 "The Inescapability of the Gettier Problem." The Philosophical Quarterly 44.

CHAPTER EIGHT: ALESSIO GEMMA

HILARY PUTNAM ON THE DIALECTICAL RELATION BETWEEN TRUTH AND WARRANTED ASSERTIBILITY

In this paper, I shall develop a personal interpretation of H. Putnam's theory of truth (Internal Realism) and, more specifically, I shall concentrate on the notion of inter-theoretical relation between "truth" and "ideal warranted assertibility". In synthesis, my overall aim is to show how Putnam's theory does not take the side of the realist nor of the anti-realist but it rather tries to overcome the whole realism versus anti-realism debate.

The paper will be structured as it follows. First, I will offer a brief sketch of the realism versus anti-realism debate in relation to the notion of truth by highlighting some current positions. Then, I shall un-pack my view of Putnam's internal perspective along two related points. On the one hand, I shall argue that Putnam's notion of truth, despite what the majority of scholars have argued, is of a realist kind. Putnam –I shall argue- has never tried to reduce/identify "truth" to an epistemic concept –rational acceptability, utility, agreement, and coherence, etc. Rather, the internalist notion of truth resembles the realist representationalist account in that it conceives truth in terms of a relation mind-world and not an intra theoretical relation. On the other hand, however, I shall argue that the following important difference holds despite his concession to realism. While on the metaphysical realist's account, truth and epistemic concepts occupy two substantially different domains; on Putnam's perspective, truth and "rational acceptability" are theoretically inter-dependent notions. Finally, I will take into considerations some lines of criticism advanced against Putnam by M. Devitt and H. Field.

The background

In order to appreciate Putnam's position, it is important to have a conceptual map of the recent developments regarding the problem of truth. In this direction, let me point to six positions that –I believe- represent a wide spectrum of philosophers involved in the truth debate:

The Correspondence Theory of Truth (B. Russell) -CTT represents the most intuitive notion of truth and many philosophers still consider it as the natural way to think about truth. Although CTT can assume different forms, all versions subscribe to at least these statements: a) Truth is a substantial property of language and b) it consists in a relation of "correspondence" between bits of language (truth bearers) and bits of the world (truth donors). Thus, Putnam writes, "Before Kant almost every philosopher subscribed to the view that truth is some kind of correspondence between idea and what is the case. However puzzling the nature of correspondence may be, the naturalness of the idea is undeniable. There is a world out there; and what we say or think is true when it gets it the way it is and false when it does not correspond to the way it is.[164]

Deflationary Theories such as Minimalism (P. Horwich) or Eliminativism (F. P. Ramsey) – Deflationary theories understand truth as a semantic device and they reject the idea that truth is a substantial property. To say of a sentence p that it is true –the deflationist says- is nothing else than to re-affirm p.[165]

Relativist Theories (B. Barnes and D. Bloor) –Relativist theories reject the very idea that truth consists in a relation of correspondence between language and bits of an un-conceptualised reality and they stress on the social character of our truth-claims. In this sense, we can also say that relativism is committed to a substantial but not representational account of truth. [166]

Coherentist Theories (W. V. O. Quine) –According to a coherentist account, truth amounts to an ideal coherence between several beliefs.[167]

Pragmatic Theories (R. Rorty) –Pragmatic theories of Truth emphasize that truth cannot be distinguished from what finally pays off or what enables us to satisfy our needs and to cope with the environment. As Rorty writes,

---

[164] H. Putnam, *Reason, Truth and History* (Cambridge: CUP, 1981), p. 50.

[165] Ramsey writes, "we put it thus 'For all p, if he asserts p, p is true', then we see that the propositional function p is true is simply the same as p, as e.g. its value 'Caesar was murdered is true' is the same as' Caesar was murdered'. See, Ramsey, F.P. 'Facts and Propositions', *Proceedings of the Aristotelian Society*, Vol.7, 1927.

[166] B. Barnes and D. Bloor write, "The relativist, like everyone else, is under the necessity to sort out beliefs, accepting some rejecting others. He will naturally have preferences and these will typically coincide with those of others in his locality. The words true and false provide the idiom in which those evaluations are expressed, and the words rational and irrational will have a similar function." D. Barnes and D. Bloor, "Relativism, Rationalism and the Sociology of Knowledge" in M. Hollis and S. Luke (eds), *Rationality and Relativism* (Oxford: Basic Blackwell, 1982), p. 27.

[167] See, W. V. O. Quine, *From a Logical Point of View* (Harvard University Press, 1961).

"Those who wish to reduce objectivity to solidarity -call themselves pragmatists – do not require either metaphysics or an epistemology. They view truth as, in William James's phrase, what is good for us to believe. So they do not need an account of a relation between beliefs and objects called correspondence, nor is an account of human cognitive abilities which ensure our species capable of entering into that relation."[168]

Verification theories (M. Dummett) – The verificationist attacks the idea (which seems to follow from CTT) that sentences might have verification transcendent truth conditions. In this direction, she underlines that the truth of a sentence cannot be conceived a part from our ability to verify it or a part from our ability to conceive under what conditions we would accept the sentence under issue. As M. Dummett writes, "I characterise as the belief that statements of the disputed class possess an objective truth-value, independently of our means of knowing it: they are true or false in virtue of a reality existing independently of us. The anti-realist opposes to this view that statements of the disputed class are to be understood only by reference to the sort of thing which we count as evidence for a statement of that class. The anti-realist insists, on the contrary, that the meanings of these statements are tied directly to what we count as evidence for them, in such a way that a statement of the disputed class, if true at all, can be true only in virtue of something we could know and which we should count as evidence for its truth."[169]

Despite the richness of this debate, the situation has reached nowadays a still position. On the one hand, the traditional and realist correspondence theory of truth, although intuitive, is doomed with several problems. Few things can be said here for obvious reasons of time. One of the problems with CTT is, for example, that it subtends a shadow of radical scepticism: namely, it makes room for the possibility that indeed even an ideal best theory could be radically false.[170] A apart from this, CTT seems also to imply beliefs that have been sharply attacked in recent time such as "the notion of passive mind", "a neo-Aristotelian view of the world", "a magical theory of reference" and so on.[171] On the other hand, however, alternative theories of

[168] R. Rorty, *Objectivity, Relativism and Truth* (Cambridge: CUP, 1991), p. 22.

[169] M. Dummett, *Truth and Other Enigmas* (Cambridge: Duckworth, 1978), p. 146.

[170] See T. Nagel, *The View from Nowhere* (New York: Oxford University Press, 1986) or H. Putnam, *Meaning and the Moral Sciences* (London: Routledge, 1978).

[171] For the relation between CTT and "the passive conception of the mind", see R. Rorty, *Philosophy and the Mirror of Nature* (Princeton: University of

truth betray our common/correspondent intuitions concerning the notion of truth for they turn truth into an internal linguistic/epistemic affair. Philosophers in the area seem to have failed, in other words, to provide an account of truth which is intuitive enough while at the same time free from at least too many philosophical problems.

Putnam and the dialectical conception of truth

As I understand it, Putnam's own view on truth is alternative to both realist and anti-realist approaches.

To start with, we need first to have a look at some relevant quotes where Putnam introduces and develops his controversial notion of truth as "warrant assertibility at the ideal limit of inquiry." Here I have selected some quotes that go from the early "Realism and Reason" in Meaning and the Moral Sciences to the late Realism with a Human Face:

Truth, in an internalist view, is some sort of (idealized) rational acceptability –some sort of ideal coherence of our beliefs with each other and with experiences as those experiences are themselves represented in our belief system – an not correspondence with mind independent or discourse-independent "states of affairs".[172]

Truth itself, on my view, is an idealisation of rational acceptability. The idea that we have a notion of truth which is totally independent of our idea of rational acceptability just seems untenable to me. [173]

[…] Kant is best read as proposing for the first time what I call the internalist or internal realist view of truth. But what is truth if it is not correspondence to the way things are in themselves? As I have said, the only answer that one can extract from Kant's writing is this: a piece of knowledge (i.e., a true statement) is a statement that a rational being would accept on sufficient experience of the kind that is actually possible for beings with our nature to have. "Truth" in any other sense is inaccessible to us and unconceivable by us. Truth is ultimate goodness to fit.[174]

The metaphysical realist insists that a mysterious relation of correspondence is what makes reference and truth possible; the internal realist, by contrast, is willing to think reference as internal to text (or

---

Princeton Press, 1979); for the relation between CTT and *essentialism* and the *magical theory of reference*, see H. Putnam, *Reason, Truth and History* (Cambridge University press, 1981) and *Realism and Reason* (Cambridge: Cambridge University Press, 1983).

[172] H. Putnam, *Meaning and the Moral Sciences*, p. 125.

[173] H. Putnam, *Philosophical Papers Vol. 3*, p. 200.

[174] H. Putnam, *Reason, Truth and History*, pp. 60-64.

theories), provided that we recognise that there are better and worst texts. "Better and "worst" may themselves depend on our historical situation and our purposes; there is no notion of God's Eye View of Truth here. But the notion of a right (or at least better) answer to a question is subject to two constrains: (1) Rightness is not subjective. What is better and what worst to say about most questions of real human concern is not just a matter of opinion. [...] (2) Rightness goes beyond justification.[175]

These passages contain the basic insights underlining Putnam's analysis of truth. However, they suggest different things. On the one hand, they clearly point that Putnam rejects what we have called a radical non-epistemic notion of truth. More particularly, they suggest that Putnam refuses both: 1) that truth consists in some sort of mirroring-correspondence between our words/sentences and some mind independent objects/facts; and 2) that the notion of "truth" makes sense a part from our notion of "ideal justification" or "ideal warranted acceptability." On the other hand, however, the above quotes suggest that Putnam rejects some natural opponents of the realist notion of truth: more specifically, the relativist approach. Where –we need to ask now- does Putnam really stand?

Many scholars have believed that Putnam does assume an epistemic notion of truth and that, specifically, he goes on reducing truth to "warrant assertibility" or "goodness to fit" or "coherence".[176] Now, although a literary reading of some of the above quotes might suggest this position, I think this would be a serious misunderstanding of the internal position. As Putnam says, in fact, in Representation and Reality:

The point is that I am not offering a reductive account of truth, in any sense (nor of warrant, for that mater). In Reason, Truth and History I explained the idea thus: "truth is idealised rational acceptability." This formulation was taken by many as meaning that "rational acceptability" (and the notion of better and worse epistemic situation", which I also employed) is supposed (by me) to be more basic that "truth"; that I was offering a reduction of truth to epistemic notions. Nothing was farther from my intention. The suggestion is simply that truth and rational acceptability are interdependent notions. Unfortunately, in Reason, Truth and History I gave examples of only one side of the interdependence: examples of the way truth depends on rational acceptability. But it seems clear to me that the

---

[175]H. Putnam, *Realism with a Human Face* (Cambridge: Harvard University Press, 1990), p. 114.

[176] See section 4.

dependence goes both ways: whether an epistemic situation is any good or not typically depends on whether many different statements are true.[177]

The pivotal concept here is the notion of interdependence. Putnam's real point is not that truth is equal to "rational acceptability" but that truth stands in a dialectical/conceptual relation to other epistemic concepts (coherence, utility, fruitfulness, etc.) and in general with the notion of "rational acceptability". Let me explain this more carefully. What Putnam is pointing at is –so to speak- a kind of twofold process. On the one hand, –he argues- we cannot understand a statements p being true apart from thinking under which rational/epistemic conditions we -or other intelligent creatures in better epistemic situations- would accept p as a true or false statement. On the other hand, we take for granted that the reason why we hold certain epistemic procedures/values as "good epistemic values/procedures" is that they lead us to true beliefs.[178] Within an internal perspective, truth and rational acceptability represent therefore a dialectical duality rather than a substantial opposition: a net of multiple conceptual relations that cannot be broken at any particular point. As Putnam writes, "This does not mean that the internal realist abandons the distinction between truth and justification, but that truth (idealised justification) is something we grasp as we grasp any other concept, via a (largely implicit) understanding of the factors that make it rationally acceptable to say that something is true."[179]

If what I have been saying is correct, then, Putnam's view is at odd with both anti-realism and realism. But, we might wonder now, if Putnam does agree that truth is not reducible to warranted acceptability but only dialectical related, how does his theory differ from the traditional conception of correspondence mind –world. Although Putnam is not clear on this point, I believe an internal realist would be happy to concede that the truth-relation in

---

[177] H. Putnam, *Representation and Reality* (Cambridge: Massachusetts Institute of Technology Press, 1988), p. 115. For similar remarks on the inter-theoretical connection between truth and epistemic concepts, see also *Pragmatism* (Cambridge: Routledge, 1995).

[178] By "epistemic tool", I mean "the ensemble of our cognitive capacities: i.e., our rational and sensorial equipment." By "epistemic procedure", I mean "the kind of procedures we use in order to acquire information about a particular object of field of study." In this sense, things like "laboratory experiments", "mathematical proofs", "logical/analytical strategies" and the like are all epistemic procedures. Finally, "epistemic values" are instead those values –like coherence, simplicity, consistency, fruitfulness etc., - which our theories are expected to meet in order to be taken into consideration.

[179] H. Putnam, *Reason, Truth and History*, p. 122.

question has a representational aspect. Let me explain. Putnam –we have seen- argues against the idea of truth as correspondence between the "mind" and the "world as it is". However, as P. Alston has noticed, it is also true that he sometimes seems to think at truth in terms of a correspondence relation between world and language.[180] This should not confuse us. What Putnam refuses -I suggest- is the idea that we can have a correspondence between bits of language and bits of an un-conceptualised world but not the idea of representation in general. In other words, Putnam is against the metaphysical conception of correspondent truth but not against the possibility of a representation between the world as represented by schemes and language. This point is particularly evident, to take an example, when Putnam talks about the possibility of having "alternative descriptions of the same facts".[181] More specifically, when he argues that both the identity and the number of objects in determinate contexts is a function of "the scheme we select" and of "the way the world is". Suppose –Putnam argues- Paul asks Tim to say how many objects there are on his table at a time t. How could Tim sort an answer out? Putnam's answer to such a question goes on two directions. On the one side, he urges us to realise that questions like the one above can be answered only from the within of conceptual schemes or frameworks. For –the argument goes- several conceptual schemes/ways of counting might give different valid answers to the same questions. However -he stresses- once we have selected a version/scheme, it is important to say that the answer to questions such, "how many object are there on the table?" is no more conventional. In other words, once we have selected our schemes, the answer is determined by the "correspondence between the way the world is according to the schemes and our sentence".

To conclude, one thing remains firm for the internal realist. Truth –although it is not an epistemic concept- cannot be understood a part from its link with epistemic concepts. And, otherwise, the reliability of epistemic concepts needs to be addressed by taking into account the realist notion of truth. The internal realist is inviting us to look at the several epistemic and semantic notions we use -when we talk about the truth of a statement- in a dialectical way. That is, she is inviting us to realise that the very notion of

---

[180] W. P. Alston, "What Metaphysical Realism is not", in W. P. Alston (eds.) in *Realism and anti-Realism*, (Ithaca: Cornell University Press, 2003), pp. 109-110.

[181] Putnam's discussion of these topic (conceptual relativity) criss-crosses his analysis of another important topic: the notion of equivalence. For further reference, see H. Putnam, *The Many Faces of Realism* (especially Lecture 1 and 2) and *Realism with a Human Face*, (especially chapter 6, "Truth and Convention").

"truth" makes sense only when we embed it into a rich environment of epistemic notions and rational practices.

Objections: Devitt and Field

Let us pause here and consider some objections. A familiar criticism against Putnam's argument is that we might think indeed at situations in which a statement is perfectly justified although plainly false and so that Putnam's idea according to which a theory which is ideally justified cannot be false is wrong. As M. Devitt writes:

> Consider the ideal theory T1. Its key features are that it is a correct predictor and that it meets all operational constrains: it fits all the possible observational evidence we could gather. From a realist perspective, to say T1 is ideal is therefore to say something about the relationship between T1, us and independent reality: our sense organs and reality are such that we could never observe anything about reality inconsistent with T1. Being ideal is species relative. To suppose that T1 might be false is simply to suppose that there might be features of reality that could not affect us at all or could not affect us in such a way that we come to right view about them. What we can observe about the world depends not only on the world but also on us and our relationship to that world. Imagine an intelligent organism like a human but with very inferior sense organs. We have no difficult in understanding how a theory might be ideal for that organism and yet from our superior perspective be clearly false. What the realist believes is that this might be so for any organism including a human being; for the realist believes reality is altogether independent of experience."[182]

I think that this criticism misses the internalist point. For one thing, it suggests that Putnam is somehow committed to a realist/optimistic theory of knowledge. However, we have seen so far that internal realism is not equal to epistemic realism.[183] The internal realist does not hold that we –human

[182] M. Devitt, "Realism and the Renegade Putnam", *Nous*, 7 (1983): p. 297

[183] By *epistemic realism*, I mean the view according to which our epistemic equipments provide us with adequate knowledge of the world (where "adequate" means, "knowledge of the way the world is independently of us"). As M. Williams writes, "Let epistemic realism be the view that in thought we may have knowledge of this mind-independent reality and that in utterance we may truly speak of it. […]In contradiction to scepticism the proponent of epistemic realism holds that we have the capacity to reflect (something of) the structure of the world in the structure of the thought. That is to say, we actually possess (some) adequate concepts. See J. Haldane, "The Mind-World Identity Theory and the anti-Realist Challenge" in J. Haldane

beings- are in some sort of representational relation to "the way the world is" thanks to the reliability of our epistemic equipment. What she does say is, rather, that "truth" and other "epistemic concepts" are inter-dependent although irreducible to each other and that, once we assume this interdependency, we can even talk about truth in terms of representation. Now, does Devitt argument provide a challenge against this idea? Does it provide an argument for the notion of verification transcendent truth condition? Well, Devitt asks us to imagine two human beings – let's call tem Tim and Paul- who are intelligent at the same degree but who have different sensual experiences. More specifically, he asks us to imagine that –say- Tim's sense organs are inferior –less complex- to Paul's ones and so that what seems true from Tim's viewpoint might indeed be false from Paul's viewpoint. Surely, Devitt's example points that the truth of a statement is independent of local epistemic means of verification: that the truth of a statement does not depend on our epistemic capacitates. The truth of Tim's theory – in the example- is in fact independent of his ability to verify or to justify it. However, the point is, can we imagine a statement being true a part from the possibility of verifying/justifying it and so on at an ideal limit? Can we really disjoint truth from rational acceptability? The answer seems negative to me. Suppose that at one point, Tim and Paul meet each other. Suppose that Paul tells Tim that his best theory was false indeed. Now, on what ground would Paul tell Tim that his theory is false? Moreover, on what ground would Tim believe and understand John's claims? I imagine that Paul would say, "Look Tim, the nature of your sense organs forces you to hold erroneous beliefs about the world. However, once you consider the statements of your theory from my perspective –on the light of my experiences-, you may realise how they turn to be false." In other words, Paul would appeal to epistemic reasons in order to make sense of the fact that Tim's theory is false and that his theory is true. He would say, among other things, that, once seen from a better epistemic position, Tim's theory shows its fallacy and that if Tim saw things from Paul's epistemic ground he could understand. However, this clearly backs up the internal point. For, what the internalist says is not that a statement that is justified according to our own capacities must –ipso facto- be true. What she simply says is that we cannot make sense of a statement p being true but completely unjustifiable. We cannot make sense of a statement being truth without imagining how it could be in principle justified by beings in good epistemic situations.

---

and C. Wright (eds.) *Reality, Representation and Projection* (New York: Oxford University Press, 1993), p. 16.

Another line of criticism I want to consider comes from H. Field. As Field writes in his critical review of Reason, Truth and History:

> Putnam goes on to tell us that truth, in an internalist view, is some sort of idealised rational acceptability; it is not correspondence. The dichotomy of idealised rational acceptability vs. correspondence is of course not exhaustive: a redundancy theorist would not accept the identification of truth with idealised rational acceptability than would a correspondence theorist. And there seem to be a good reason to reject the equation of truth with idealised rational acceptability. For the number of individual dinosaurs that have ever existed is surely no more than $10^{10\ 10}$. Consequently, either exactly zero dinosaurs have existed or exactly one dinosaur has existed, or, .., or exactly $10^{10\ 10}$ dinosaurs have existed. Consequently, assuming the Tarski schema "p is true if and only if p", exactly one of the following statements is true: S0 There have been exactly zero dinosaurs; S1 There have been exactly one dinosaur; ... $S10^{10\ 10}$ There have been exactly $10^{10\ 10}$ dinosaurs. It follows that if truth is rational acceptability, exactly one of those $10^{10\ 10}+1$ statements is rationally acceptable. But that is absurd: the disjunction is rationally acceptable, but clearly no evidence we now have or ever will have will make exactly one of these statements rationally acceptable. [...] Putnam claims that on any theory of truth "Snow is white" is equivalent to "Snow is white" is truth, but his own theory of truth (truth= idealised rational acceptability) seems to be an exception.[184]

Two points- I suggest- go wrong about Field's argument. The first one is that Field misinterprets Putnam when he says that an internal realist is committed to a reductivist notion of truth. As I tried to make clear, in fact, Putnam does not attempt to reduce truth to an epistemic concept –warrant acceptability, idealised coherence, utility and the like. What he says is rather that "truth" and "epistemic concepts" –though not reducible to each other- are conceptually inter-dependent and therefore there is no room for the notion of verification-transcendent truth-conditions.

The second point regards, instead, the connection between Internal Realism and Tarski's Schema. Field reminds us that Tarski's T-Schema –p is true if and only if p- must obtain on every theory of truth. However, he goes on, internal realism represents an exception to this rule, for – on such view- we have that "p is y" turns to be not equivalent to "p is y" is true –where "truth" is equal to "rationally acceptable". Now, I think that Field's example begs the internal realist question for the following reason. Field first asks us to agree that the numbers of dinosaurs that have existed in the earth is surely no more than $10^{10\ 10}$. Therefore, he keeps on, if we ask, "How many dinosaurs have existed?" the true answer must give us a value that goes

---

[184] H. Field, "Realism and Relativism" in *The Journal of Philosophy*, 79, (1982): p. 556.

between 0 and $10^{10\ 10}$. That is, the correct answer will be one of the following: "there have been exactly zero dinosaurs" or "there has been exactly one dinosaur" and so on until our ideal number of $10^{10\ 10}$ dinosaurs. However, –the argument goes- if we take "truth" to be equal to "rational acceptability" we should admit that only one and one statement of the above series is indeed rationally acceptable. In fact, since there is only one true answer to our question, if truth is rational acceptability then there has to be only one rational acceptable answer. Nevertheless, Field concludes, this conclusion is clearly unintelligible for, while the disjunction of our series of answer is rationally acceptable, we have no mean to pick one and one statement as the only rationally acceptable. Now, the problem is that Field's premise (the question "how many dinosaurs have ever existed?" allows for only one and one true answer) rests on the adoption of a realist essentialist picture of natural kind terms. More particularly, it rests on the idea that terms such "dinosaurs" have fixed boundary and that such boundaries are fixed by nature in itself. In other words, Field's example rests on the assumption of what Putnam has often labelled as a ready-made picture of the world or a world with a built in structure. However, in assuming essentialism as a starting point for his argument, Field begs the question of the internal realist. For, an internal realist would say that –pace essentialism- the above question allows for different answers depending on the taxonomic system we decide to us. More particularly, the internalist would say that the term "dinosaur" is highly conventional and that it is a matter of debate whether certain boundaries entities have to be accounted as dinosaurs or not.[185] Therefore, she would conclude, Field's question allows for different answers or, better, Field's question –so as it stands- is unanswerable.

---

[185] That a question like "how many dinosaurs have ever existed?" has not a univocal answer should not sound strange. Indeed contemporary science offers us plenty of reasons to dismiss such absolutistic questions as meaningless. To report to just one example, there has been a great talking recently about the discovery of a new planet –Sedna. However, although the scientific community acknowledges the discovery of Sedna as an important step toward a better understanding of our solar system-, there is no agreement on whether Edna is or is not a planet. In this direction, what the majority of scientists say is that Sedna is a planet under a certain description and it is not a planet under others. This situation points that even a question like, "how many planets are there in our solar system?" does not allow for a univocal answer and this –it is important to repeat- it is not due to lack of knowledge but to the elasticity of our taxonomical systems.

Conclusion

In this paper, I have analysed Putnam's internal realism in relation to a main problem: the relation between "truth" and "justification". In this direction, we have seen how Putnam's thinking underlines a common feature that stands at the very core of his internal realism: an anti-dualist frame of thought. That is, the movement from an essentialist way of thinking to a dialectical one or from considering concepts in isolation to analysing them in their multiple conceptual relations. This reading has led us to discharge some common interpretations of Putnam's philosophy that make it indistinguishable from various anti-realist proposals such as idealism or relativism. Internal Realism, I have stressed, is at odd to both realism and anti-realism in that it rejects what these philosophical currents accept: dualism, reductionism, essentialism. Internal realism rejects the possibility that we can ask questions about –say- the "mind" unless we take into account important beliefs about the "world" or –again- that we can make any sense of "a statement being true" apart from "understanding under which possible conditions an intelligent being would justify it" and so on. Internal realism is then -at its bottom-, an anti-Socratic philosophy for it cuts at its roots our impulse to ask the "what is it?" questions by replacing them with a holistic and dynamic approach.

Bibliography

Alston, W. (2003), "What Metaphysical Realism is Not", in Alston, W. (ed) (2003), *Realism and Anti-Realism* (Ithaca: Cornell University Press).

Barnes B. and D. Bloor, (1982), "Relativism, Rationalism and the Sociology of Knowledge" in M. Hollis and S. Luke, (1982), *Rationality and Relativism* (Oxford: Basic Blackwell)

Devitt, M. (1983), "Realism and the Renegade Putnam: A Critical Study of Meaning and the Moral Sciences" in *Nous*, 7: pp. 291-301.

Dummett M., (1987), *Truth and Other Enigmas* (Cambridge: Duckworth)

Field H., (1982), "Realism and Relativism" in *The Journal of Philosophy*, pp. 552-567

Goodman N., (1978), *The Ways of World Making* (Indianapolis: Hackett Publishing Company).

Haldane, J. (2003), "Mind-World Identity Theory and the Anti-Realist Challenge" in Haldane J. and Wright C. (1993) (eds.), *Reality, Representation and Projection* (New York: OUP)

Nagel T., (1986), *The View from Nowhere* (New York: Oxford University Press)

Putnam, H. (1975a), *Mathematics, Matter and Method* (Cambridge: Cambridge University Press).

———. (1975b), *Mind, Language and Reality* (Cambridge: Cambridge University Press).

———. (1978), *Meaning and the Moral Sciences* (London: Routledge & Kegan Paul).

———. (1981), *Reason, Truth and History* (Cambridge: Cambridge University Press)

———. (1983), *Realism and Reason* (Cambridge: Cambridge University Press).

———. (1987), *The Many Faces of Realism* (La Salle: Open Court, 1987).

———. (1988), *Representation and Reality* (Cambridge: Massachusetts Institute of Technology Press).

———. (1990), *Realism with a Human Face* (Cambridge: Harvard University Press).

———. (1992), *Renewing Philosophy* (Cambridge: Harvard University Press).

———. (1995), *Pragmatism* (Cambridge: Blackwell).

Ramsey, F.P. (1927) 'Facts and Propositions', *Proceedings of the Aristotelian Society*, Vol.7

Rorty R., (1979), *Philosophy and the Mirror of Nature* (Princeton: University of Princeton Press)

———. (1991), *Objectivity, Relativism and Truth* (Cambridge: CUP)

Williams M., (1984), "Review of *Reason, Truth and* History" in *The Journal of Philosophy* 81: pp. 257-261.

INDEX

ABOUT THE EDITORS

Barbara Gabriella Renzi is completing her PhD on Scientific Methodology and Evolutionary Biology at the Queen's University of Belfast. She studied Philosophy of Science and specialised in Bioethics at the university 'La Sapienza' in Rome. She recently published "Natural Selection, Exaptation and the Evolution of Science" in *Philosophical Writings* and her poems appear in Italian anthologies.

Stephen Rainey is a research student in the School of Philosophical Studies at Queen's University Belfast. He obtained his BA Philosophy First Class Honours and MA Philosophy with Distinction at the same institution in July 2003 and December 2004 respectively.